Biomass as Fuel

L. P. WHITE

*General Technology Systems Ltd
Forge House
20 Market Place
Brentford
Middlesex*

and

L. G. PLASKETT

*Biotechnical Processes Ltd
Hillsborough House
Ashley
Nr. Tiverton
Devon*

ACADEMIC PRESS   1981
A Subsidiary of Harcourt Brace Jovanovich, Publishers
London   New York   Toronto   Sydney   San Franciso

ACADEMIC PRESS INC. (LONDON) LTD
24/28 Oval Road
London NW1 7DX

United States Edition published by

ACADEMIC PRESS INC.
111 Fifth Avenue
New York, New York 10003

Copyright ©1981 by
ACADEMIC PRESS INC. (LONDON) LTD

All Right Reserved

No part of this book may be reproduced in any form by photostat, microfilm, or any other means, without written permission from the publishers.

British Library Cataloguing in Publication Data

I. White, L. P. II. Plaskett, L. G.
   Biomass as fuel.
   1. Biomass energy
   I. Title
   662'.6   TP360
   ISBN 0-12-746980-X
   LCCCN 81-66689

Printed in Great Britain by The Alden Press (London & Northampton) Ltd

Preface

Surprisingly to most people in developed countries about one seventh of the energy used throughout the world at the moment comes from firewood and firewood is the most important fuel for the bulk of the world's population. This is because it is usually the only domestic fuel available to the populations of the third world. This book is not, however, primarily about firewood, but rather the largely unrealised potential for fuel and energy represented by other forms of vegetation and animal matter that are broadly termed biomass.

In approaching these, however, the ancient technologies and economies of wood burning provide useful precedents and examples at several points against which to compare the newer propositions. Many wood burning applications and some other biomass energy practices were wiped out by the huge extension of the use of more convenient fossil fuels like oil. This trend now shows every sign of going into reverse with inevitable long-term rises in all fuel costs in real terms. In the Third World in particular there are signs of an actual flight from imported oil which has increased in real cost five fold within the decade. To cope with this situation a whole range of strategies is being adopted including energy conservation (which, by definition, is limited) and the use of alternative fuels to stretch fossil fuel supplies. The use of biomass energy is one component of the second strategy and is particularly attractive in that, unlike fossil fuels, biomass is renewable.

The impetus and urgency of the world energy problem has justified the examination of a very large number of propositions for the use of renewable biomass fuels. The subject as a whole involves a wide range and mixture of high and low technologies, extensions of well-known ideas and revolutionary concepts that could heavily affect not only the existing energy supply patterns but also the industries with which biomass is associated, like agriculture and forestry. Similarly some propositions will have major impacts on land use and employment. At the other extreme some prospects call for no more than individual initiatives in low cost developments.

National and international programmes of research and development cover a wide range of prospects and other research in scientific and engineering disciplines associated with biomass has given rise to an extensive literature. However, up to now there have been few publications which give any sort of comprehensive coverage to the subject in all its lines of development, and in any depth. It is the aim of this book to fill this need and to set out the relative feasibility, time-scales and overall potential of the welter of ideas. It is aimed firstly at post-graduates and undergraduates in appropriate subjects, but also at a more general, non-specialist readership concerned with energy strategy and economy, industrial and land development.

As the individual subject areas are so interlocked it should also serve as a useful reference work for specialists in particular fields. It requires no more than a basic grounding in scientific or engineering concepts and terms for the ideas in it to be understood.

Essentially the book is organised to review the six main categories of prospect for using biomass as an energy or fuel source, covering both waste materials and the ideas for energy plantations, using examples of particular routes and their problems. Prior to this, however, there is an introduction to the basic principles of energy capture from the ultimate source — sunlight — and its incorporation into living tissues, and to the technologies of biomass fuel use and conversion which are common to all concepts. Thus the basic nature of biomass material and its capacity to convert, store and release energy is outlined in the light of basic efficiency and practical limitations. The chemical and energy characteristics of the various materials are described and the thermal, chemical and biological conversions that are used ultimately to provide heat or fuel. The special aspects of these that are relevant to particular routes are dealt with also in the appropriate self-contained chapters.

In a final chapter the question of the importance of biomass energy is appraised in the light of the total potential, overall energy strategies, and the main directions and thrust of development.

May 1981

L.P.W.
L.G.P.

Contents

Preface		v
1	**Biomass**	1
2	**Fuel use and conversion technologies**	11
	A Nature of biomass as a fuel: Advantages and disadvantages	11
	A1 Purposes of conversion processes, 17	
	B Types of conversion process	19
	C Direct combustion	19
	D Thermal processing	23

    D1 The nature, objectives and advantages of thermal processing, 23
    D2 Different versions of the process and conditions of reaction, 25
    D3 Properties of biomass relevant to thermal processing, 29
    D4 Preparation of biomass for thermal processing, 29
    D5 Types of thermal processing equipment, 30
    D6 Gas cleaning, 32
    D7 Utilisation of pyrolysis or gasifier gas, 32

	E Biological conversion processes	34

    E1 Anaerobic digestion, 34
    E2 Production of ethanol (ethyl alcohol), 43
    E3 Other potentially available bio-conversion processes relevant to fuel applications, 46

3	**Crop wastes**	49
	A Range of sources and availability of crop wastes	49

    A1 Residues dry enough for combustion, 49
    A2 Wet agricultural residues, 51

	B Quantities and collectability of crop wastes	52

	B1	Cereal straw, 52	
	B2	Sugar beet tops, 53	
	B3	Potato haulm and reject tubers, 54	
	B4	Haulms and shucks from peas and beans, 55	
	B5	Brassica wastes, 56	
	B6	Wastes from other vegetables, 57	
	B7	Woody wastes in agriculture, 57	

    C Logistics of collection and storage: Geographical density of arisings   58

    D The composition of crop residues   62

    E Methods of utilising crop residues for energy   66

    F Competitive uses of potential crop waste as energy feedstocks   68

        F1 General, 68
        F2 Straw and other dry residues, 69
        F3 Competitive uses for wet agricultural residues, 72

4 Animal wastes   **73**

    A Characteristics of animal wastes   73

        A1 Availability of animal wastes, 75
        A2 Storage of animal wastes, 77
        A3 Quantities of animal wastes and gas production potential, 78

    B Animal bedding   80

    C Sizes of livestock unit justifying a digester installation   81

        C1 Cattle herd sizes, 84

    D Competitive uses for livestock wastes   84

    E New approaches to utilisation of animal wastes: Solid/liquid separation   85

5 Energy farming: Natural vegetation and dedicated energy crops   **89**

    A The concept of energy farming   89

    B Annual agricultural crops as dedicated energy crops   91

    C Catch crops   95

    D Perennial non-woody plantation crops   99

	E	Processing options for non-woody biomass: Integrated feed/fuel production systems	104
	F	Harvesting natural vegetation	108
	G	Summary of productivities and costs	110
6	**Wood and wood wastes**		**111**
	A	Firewood	111
	B	Wood	111
	C	Wood waste	113
	D	Direct combustion	116
	E	Pyrolysis and gasification	118

        E1 Wood gas, 119
        E2 Methanol, 120
        E3 Pyrolysis oil, 121

	F	Other processes	121
	G	Development considerations	122
7	**Short rotation forestry – SRF**		**123**
	A	Why short rotation?	123
	B	Yields	124
	C	SRF in practice	127

        C1 General factors, 128
        C2 Energy balance, 129
        C3 Land available, 129

	D	Research and trials	130
	E	Viability	131
8	**Sewage and municipal wastes**		**135**
	A	Sewage	135

        A1 Nature and quantity of arisings, 135
        A2 Methods of sewage treatment, 137
        A3 Digestion of sewage sludge, 139
        A4 Economic considerations, 140

	B	Municipal solid wastes – MSW	141

        B1 Direct combustion, 144
        B2 Refuse-derived solid fuel – RDSF, 145

	B3 Digestion, 147	
	B4 Land-fill gas, 148	
	B5 Pyrolysis and gasification of MSW, 151	
	B6 Economic considerations, 152	
	B7 Hydrolysis and fermentation, 153	

9 Aquaculture and marine harvesting — 155

A Freshwater macrophytes — 155

B Seaweed — 157

- B1 Stocks, 160
- B2 Mariculture, 160
- B3 Fuel generation, 164
- B4 Alternative uses, 164
- B5 Prospects overall, 166

C Microalgae — 167

10 Overall prospects and limitations — 175

A Possibilities — 176

- A1 The methane economy, 176
- A2 Methanol, 177
- A3 Ethanol – gasohol, 178
- A4 The hydrogen economy, 179
- A5 Solutions in perspective, 179

B National programmes — 181

- B1 Brazil, 181
- B2 The United States, 182
- B3 Canada, 183
- B4 Europe, 183
- B5 Developing countries, 184
- B6 Institutional and political constraints, 184

Units and abbreviations — 189

References — 191

Subject Index — 205

[1]
Biomass

All organic matter, or biomass, can in one way or another be used as fuel. It is composed mainly of carbohydrate compounds the building blocks of which are the elements carbon, hydrogen and oxygen. All ultimately derive from the process of photosynthesis in plants, but may be in many forms, vegetable or animal. Biomass from which energy can be reclaimed can be harvested as grown crops or natural stands for energy supply, or as surpluses or wastes from crops grown primarily for food and manufacturing raw materials, and through municipal and industrial waste.

Vegetable biomass in the form of wood or dry litter was the first, and for a long time, the universal and only fuel providing external energy for cooking, warmth, and later, manufacturing and until recently it was generally abundant. With the world population growth and the development of industrial demand supplies have become increasingly scarce notwithstanding the switching of the developed world to fossil fuels with their higher energy densities and convenience. Due to such factors as over-grazing and deforrestation fuel supplies from biomass are now in a critical state in many under-developed countries in which the mass of the population relies on them. In developed countries, however, their potential has been largely ignored until recently when the pressures on fossil fuel supplies are also seen as approaching crisis proportions.

Of the fossil fuels, coal was in local use in the Middle Ages but did not begin to be exploited in quantity until the nineteenth century. The use of oil is predominantly a phenomenon of this century and is likely to die out in the next. These fuels upon which our current civilisation is largely based, are non-renewable and the limits to their availability are clearly seen. It is for this reason that so much attention is now being given to the search for alternative and supplementary fuel supplies including biomass, which has the advantage of being renewable. It shares this characteristic with hydro-electric power, wind and wave power, and the direct use of solar energy. Also, even in its large-scale exploitation it is likely to be relatively non-polluting. However, as will be seen later, its environmental impacts, if adopted on a large scale, could be very considerable, particularly in terms of land use changes.

Energy from biomass is therefore attractive on a number of counts, including the claim that its primary source — sunlight is free. However, there are basic limitations both in the generation of the energy feedstocks — wood or other plant matter — and in their use of conversion to other more conveniently handled fuels like alcohols, pyrolytic oil or gas. The fundamental drawback is the commitment to collect the primary energy input from the Sun, which is at a relatively low power density, using rather inefficient collectors — that is plants; for instance field crops on temperate zones really only work at efficiencies of 1 percent or less over a season.

Looking at this in terms of potential energy supply; the whole solar energy receipt over the 230 000 square kilometres or so of the United Kingdom, at an average rate of 3300 MJ/m^2/yr would total 759 × 10^9 GJ/yr. Putting this into the more manageable unit of million tonnes oil equivalent (1 Mtoe = 44 × 10^6 GJ) the annual solar energy receipt is approximately 17 250 Mtoe. This compares to a total energy content of all fuel used — coal, oil gas, nuclear — of about 300 Mtoe in 1978. However, if only 1 percent of the solar energy is fixed by plant tissues, even if all land was devoted to growing crops primarily to supply energy, only 57 percent of the 1978 consumption could be supplied. At a more realistic average of 0.4 percent efficiency, the contribution would be 23 percent.

Many versions of these hypothetical calculations can be devised, substituting different solar energy capture efficiencies and energy use predictions. Always they indicate that even large-scale energy from biomass schemes can only make limited contributions to the energy needs of most developed countries. However, compared to other alternative energy approaches, there are fewer technological problems. The chief obstacles to the many and varied opportunities for providing energy in this way are in the economics of the proposed schemes.

Potential sources of biomass that can be used directly for energy or converted to more convenient transportable (liquid) fuels or transmitted energy (electricity) range from dry crop wastes like straw to wet slurries of animal waste; from sawdust to crops grown specifically for their energy content; and from seaweed washed up on the shore to tanks of cultured microalgae. Each of these potential energy feedstocks has its special problems in cultivation, collection or concentration and the development of the techniques for these are at a variety of different phases. Some techniques can be seen to be economic at present, others are still experimental. The same is true of the technologies of conversion to more useful forms of energy which range from burning to generate heat directly; digestion or fermentation to produce gas or alcohol; or thermal processes to produce gas and liquid fuels. The possibilities vary with the different feedstocks and involve different energy economies based on different efficiencies of conversion, which affect the economic feasibilities of the schemes as a whole.

The economic feasibility of any energy from biomass scheme depends ultimately on the cost of competitive conventional fuels. Apart from this their practical realisation, if technically feasible, will be influenced by the value of the feedstock as determined by the demand for it for other uses. For example some straw may have a higher value as bedding for animals. Some schemes appear to make sense only on a small scale and when integrated with existing activities like farming. Others like the proposed large-scale forest energy plantations will involve major land-use changes with large social and environmental impacts. Some types of farm-scale scheme can be effectively subsidised by the extensive use of standard equipment normally used for the business of agriculture. Others will require major inputs in the form of specialised equipment and the adoption of new cultivation practices. In a sense then, the scale also affects the feasibility, smaller scale schemes being easier to implement; but on the other hand only large schemes may achieve the economies of scale to make them viable.

The amount of tissue that a plant produces depends firstly on the amount of solar energy the plant receives and can store in carbohydrates. The solar radiation received by the earth at the outer limit of the atmosphere directly facing the Sun is at the more or less steady level of 1353 watts per square centimetre. This is the Solar Constant. Over a year this totals 10^9 watt/seconds per square metre — about 42.6 gigajoules per square metre per year. However, the Earth is a rotating globe and this highest intensity is only received directly facing the Sun.

Full solar zenith (the Sun directly vertical overhead) only occurs in the Tropics. Elsewhere, and for most of the time in the Tropics, the peak energy intensity is dispersed over an effectively larger ground areas. Further, much of the solar radiation is lost in that about 42 percent is reflected back by the Earth's surface, and clouds, and more is absorbed by the atmosphere.

What actually arises at the surface of the Earth, and can potentially be used by plants after passing through the atmosphere depends, therefore, on the Sun's range of elevation at different times of the year and the weather and the state of the atmosphere. The amount of energy the plant receives also depends on day length — which, like solar elevation, is a function of season and latitude. The state and efficiency of the plant canopy and local conditions of shading are important in individual cases.

The intensity of the photosynthetically active radiation — corresponding approximately to visible light (which is at wavelengths of 400 to 700 nm) — at solar zenith is usually in the order of 800–1000 W/m^2. With all the other factors in play the actual solar energy receipts, as shown by the measured values for representative sites at different latitudes and with different climates in Table I, vary between 2.5 to 8 GJ/m^2/yr.

These are the quantities of energy potentially available to plants. However, in the build-up of biomass by the process of photosynthesis, overall only some 0.1 percent of the total average solar radiation energy received at the surface of the Earth is actually captured by plants. This is partly, of course, because not all of the Earth's surface is covered by growing plants all of the time, but also because of the low energy conversion efficiency of the photosynthetic process in individual plants.

Table I Total solar energy received at different sites throughout the world

Site	Latitude	Region	MJ/m² /yr
Reykjavik: Iceland	64 °48N	North Atlantic	2779
Barrow: Alaska	71 °18N	Sub-polar	3206
Kew: London	51 °28N	Temperate Western Europe	3397
Valentia: Ireland	51 °56N	Temperate Western Europe	3735
Paris: France	48 °49N	Temperate Western Europe	4053
New York: USA	40 °47N	Eastern Continental	4947
Nice: France	43 °42N	Mediterranean	5580
Manáus: Brazil	3 °08S	Equatorial forest	6169
Bombay: India	18 °56N	Tropical–Monsoonal	7558
Khartoum: Sudan	13 °56N	Dry tropics	7741
Phoenix: Arizona	33 °26N	Sub-tropical desert	7940

Figures variously from: World distribution of solar radiation (Lof, G.O.G., Duffie, J.A. and Smith, C.O.) *Solar Energy* No. 10, 27–37 (1966). Converted from annual mean Langleys per day. 1 ly (cal/cm²) as 4.184 J/cm². "European Solar Radiation Atlas" (Palz, W., ed.). W. Groschen Verlag, Dortmund (1979). Converted from annual mean Wh/m²/d. 1 MJ/m² as 278 Wh/m².

Essentially photosynthesis converts carbon dioxide from the atmosphere and water from the soil or growing medium to the carbohydrate substances that make up the bulk tissues and which then contain a proportion of the solar energy trapped in their chemical bonds. This conversion is carried out with the aid of the catalytic chemical chlorophyll which absorbs the electromagnetic radiation of sunlight. Green plants have evolved their energy filtration mechanism to exactly match the characteristics of the Sun as an energy emitter. There are some minor variations of this in some of the lower forms of plant life like rhodophytic bacteria which synthesise carbohydrates more efficiently using the red sector of the visible light spectrum.

Photosynthesis is the process on which all the basic food chains are based and therefore on which all life depends. It also releases oxygen and it is from this source that the present composition of the Earth's atmosphere derives and is maintained.

The amount of energy fixed by plants in this way is regarded as their Gross Primary Production (GPP). It does not all however, become available for use for secondary energy supply. Some of the energy is used by the plants themselves to grow, so the amount available in the tissues or biomass becomes the Net Primary Production (NPP). Typically in temperate grassland this would be some 75 percent of the GPP.

There are varying claims for the maximum photosynthetic efficiency achieveable but it is generally regarded as not much higher than 10 percent. This sort of level is achieved under ideal laboratory conditions. In the field, in optimised conditions, 3 to 5 percent is the probable maximum for high-yielding plants, but this is usually sustained only for short periods of peak growth.

It is unwise therefore to base an annual yield prediction for a crop on its maximum photosynthetic rate. In practice the average rate, over a growing season, for a high-yielding temperate zone crop will be 1 percent or less, though for some tropical crops it is normally higher. The reason for this is not necessarily the higher sunlight intensities in the tropics. The photosynthetic process is in fact saturated well below the highest intensity levels and above this its productivity rate falls off sharply. It is also less efficient at high temperatures with a peak around 35 to 40 °C. Naturally though, tropical plants are better adapted to higher light intensities and temperatures.

Some tropical and sub-tropical plants in fact have a different photosynthetic route — the C_4 as opposed to the C_3 process. In the C_3 process the compound initially fixed by enzyme action is 3-phosphoglyceric acid (PGA) which contains three carbon atoms. In C_4 plants, different enzymes produce the four-carbon acids, malate or aspartate. Typical C_4 plants are maize and sugar cane, the former of which at least has varieties adapted to warm temperate climates. The C_4 process can operate at higher light intensities than the C_3 and though it does not appear to be markedly more efficient than the C_3 process. C_4 plants are generally more productive because, growing mostly in lower latitudes they can better utilise the high sunlight levels prevailing. Most temperate crops like beet and potatoes are C_3 plants but not all tropical crops are C_4. Rice, for instance is a C_3 plant. Algae also are believed to follow the C_3 path.

In practice, the efficiency with which plants can use the photosynthetic process to fix carbon depends on a number of environmental factors and maximum efficiency can only occur under optimum growing conditions, with ample water, nutrients and the absence of disease. Outside the Tropics temperature is critical with plant growth only taking place above 5 °C, so that the length of the growing season is the main controlling factor on overall productivity. Though many plants can achieve very high growth rates for limited periods, these are not maintained throughout the year and cannot simply be extrapolated to calculate annual productivity rates.

Plants like sugar cane, growing continuously in near ideal conditions, have given some of the highest recorded yields. However, even in these circumstances short-term growth rates vary considerably with the cycle of plant development. Some quoted annual productivities for different crop and vegetation types are given in Table II.

In calculating the amount of energy that can be harvested in plants from a given area it is necessary, of course, to know the average energy content of the

Table II Production from terrestrial vegetation. (From Boardman and Larkum, 1975)

	Productivity (t/ha/yr)
Tropical	
Napier grass (*Pennesetum purpureum*)	88
Sugar cane	66
Reed swamp	29
Temperate	
Perennial crops	
Annual crops	22
Grassland	22
Evergreen forest	22
Deciduous forest	15
Savannah	11

plant material. Though there can be considerable variation, most air-dry plant material with about 15 percent moisture content — as opposed to totally oven-dry matter — can be assumed to contain 16–18 megajoules per kilogram (or 16 gigajoules/tonne). If this energy content is to be made available, the material has to be burnt directly, or converted by chemical or biochemical processes to fuels like gas, alcohol or oils. Each of these conversion processes results in a large loss of energy so that the available energy from the original 16 MJ/t could be only 35 percent or less than this. Energy conversions and efficiencies of use are further discussed in chapter 2.

Calculations of the maximum photosynthetic efficiency for the conversion of solar energy to biomass vary, but the theoretical maximum ranges from 8 to 11 percent (Hogg, 1971; Schneider, 1973). Most crops and natural vegetation stands are much less efficient than this. Over most of northern Europe 1 to 5 percent is the probable range for crops under normal field conditions, and represents, over one growing season, a dry biomass yield of between 20 and 30 t/ha (Long, 1977). This can be compared to a suggested maximum productivity for C_3 plants of 40 t/ha/yr dry matter (Loomis and Gerakis, 1975).

There is in fact probably little difference in the potential maximum productivity of crops, forest and other natural vegetation. For root crops like beet and potatoes 50 t/ha/yr (harvested, not dry) yield can be expected in temperate zones (Thom, 1971; Penman, 1971). In practice, production figures for potatoes in European countries vary from 16.6 to 33.8 t/ha. Newbold (1971) indicates that grain grass crops in temperate zones could give maximum yields of 10 t/ha dry matter. Penman (1971) suggests that a practical aim for these in normal conditions of farming in the United Kingdom would be 12 t/ha dry matter or 5 t/ha of grain. By comparison yield figures in 1978 for European Community countries were: Wheat: 3.0–5.5 t/ha; Barley: 2.6–5.0 t/ha; Maize: 3.0–6.57 t/ha.

Estimates for forest production vary widely and are influenced by the age of the stock as measured and by the longer growing cycles. Westlake (1963) suggests that productivity levels in temperate deciduous forests are typically 12 t/ha/yr plus or minus 25 percent, and for coniferous forests 28 t/ha/yr plus or minus 25 percent. Specific figures quoted by this author are: Birch: 8.9 dry t/ha/yr; Alder: 16.0 dry t/ha/yr; Scots Pine: 16.0 dry t/ha/yr; Grand Fir: 35.0 dry t/ha/yr.

Work in the Netherlands (Frissel *et al.*, 1978) has indicated that the maximum possible theoretical production of fast growing poplars under optimum conditions of nutrient and water supply is 44 t/ha/yr dry matter. This compares the normal annual production rate of the same species of 6 dry t/ha.

Comparable yield figures are quoted for a number of non cultivated herbaceous plants. For instance Westlake (1963) gives 26 dry t/ha/y for *Pteridium* fern — though Pearsall and Gorham quoted by Newbold (1971) give production figures for the same plant as 9.8 to 14 t/ha/yr.

From the above it is evident that yields of terrestrial biomass of 20–30 dry t/ha/yr can be expected but, except in exceptional and limited natural circumstances, only under optimum conditions equivalent in management and input levels to a high standard of agriculture. A close approach to the apparently ultimate limit of production rate giving 40 dry t/ha/yr is only possible during short periods of growth cycles and would rarely, if ever, be sustained for a full growing period to give annual yield of this magnitude.

In freshwater naturally nutrient-rich or eutrophic sites can be highly productive. Though Merril (1974) indicates typical biomass productivity in lakes and streams in North America as 5.4 (dry) t/h/y and swamps and marshes at 22 t/h/y Westlake (1963) suggests productivity from good wet sites in Europe to be in the range of 20–37 dry t/h/y — similar to high figures for terrestrial sites. More typical figures (quoted by Newbold, 1971 and Westlake, 1963) are:

Carex fen (wet grassland)	UK	4.2–6.3
Phragmites swamp	UK	7.5–13.0
Typha swamp	UK	10.7
Eutrophic lake (*Ceratophyllum demersum*)	Sweden	9
Pond (*Berula* and *Ranunculus*)	UK	8.5

Natural levels for freshwater microalgae are low. Westlake (1963) suggest 1–3 dry t/h/yr from typical lakes with 4–9 dry t/h/yr from eutrophic lakes. Production from cultured algae however, can be truly spectactular. Production rates of freshwater algae as high as dry 110 t/h/yr are claimed from Israel (Shelef *et al.*, 1976). Yields of *Scenedesmus* species from California and Thailand are of the order of 55 t/h/yr. A reason for this may be as Soeder (1976) points out, that because the nature of the aquatic environment, where plants are surrounded by their growing medium, even in non-optimised conditions some cultures of algae are able to exceed the theoretical maximum potential production of terrestrial plants in the same latitudes.

In the sea primary bio-productivity is carried out by phytoplankton or by macrophytic seaweeds. Phytoplankton is predominantly composed of free-floating microalgae. Macrophytic seaweeds, are mostly anchored algal species and their growing zones in shallow waters are essentially limited by light penetration through the water. This is true also of the sole non-algal angiosperm seaweed *Zostera* or eel-grass. There are a few examples of free-floating macrophytic algae like *Sargassum*.

The typical temperature range for phytoplankon growth is 17 to 35 °C with marked summer and autumn peaks in temperate waters due to thermocline effects. The primary control on productivity however seem to be the availability of nutrients. In general, even in upwelling areas and nutrient-rich coastal zones, productivity is low in comparison with terrestrial bio-productivity.

Productivity in temperate waters is quoted by Round (1966, after Gessener) as 200 to 400 grams of carbon per square metre a year or 2 to 5 t/h/yr total bioproduction – agreeing with Westlake (1963) who quotes 3 t/ha/yr plus or minus 50 percent. As is the case with freshwater microalgae, cultured marine species can give very high production rates.

Westlake suggests a typical productivity for temperate marine macrophytes at 29 t/ha plus or minus 15 percent and quotes production rates for two European species growing in Novia Scotia as: *Ascophyllum:* 20–26 t/ha/yr; *Laminaria* 48 t/ha/yr.

These must be wet tonnage figures. The dry tonnage production of the fastest growing species known – the *Macrocystis pyrifera* of the Californian coast is quoted by Jackson as typically 21 t/ha/yr (North, 1977, after Clendenning).

Production of the small macrophyte *Gracilaria* in experimental ponds in Florida has exceeded a rate of 112 t/h/yr dry material (Ryther *et al.*, 1977) which is the highest productivity rate so far reported. This, however, is calculated from a rate measured over a short period of optimum growth and it is not claimed that the true annual production will be anywhere near this figure.

These levels of production in terrestrial, freshwater or saltwater environments are what any practicable energy-from-biomass scheme must be based on. A certain tonnage with a certain energy content has to be harvested and, after processing, a somewhat smaller amount of energy is made available as fuel. All the processes of establishing and running the schemes absorb energy – for cultivation, transport of stock etc., and the conversion processes. Obviously to be worth doing or to be economic, the energy inputs must generally be less than the energy output as fuel so that there are identifiable minimal production levels at which any scheme is a viable proposition. Also the fuel produced must be competitive in price with equivalent competitive fuels produced from the other sources. The fact that conventional fossil fuel costs are increasing all the time is, of course, a factor in favour for looking at energy from biomass as well as a number of other options for alternative energy. However, the perceived advantages of energy from biomass are not necessarily straightforward.

[2]
Fuel use and conversion technologies

A Nature of biomass as a fuel: Advantages and disadvantages

All fossil fuels originated from biomass and one cannot help but be impressed by the major physical and chemical differences between these fuels and the original biomass from which they were derived. Present-day reserves of coal, oil and natural gas have been subjected to extremely effective cost-free processing operations through the combined action of climate and geological forces over enormous spans of time. Fresh biomass has serious disadvantages as a fuel compared with fossil fuels; the geological processing operations have been very effective in overcoming these.

The disadvantages are (*i*) that biofuels usually have only a modest thermal content compared with fossil fuels, (*ii*) they often have a high moisture content, which has the effects of inhibiting ready combustion, causing major energy loss on combustion, mainly as latent heat of steam, and also rendering the material putrifiable so that it cannot be readily stored, (*iii*) they usually have a low density and, in particular, a low bulk density, factors which increase the necessary size of equipment for handling, storage and burning, (*iv*) the physical form is rarely homogeneous and free flowing, which militates against automatic feeding to combustion plant. The geological conversion processes which formed fossil fuels have increased the thermal content per unit weight relative to fresh biomass, virtually eliminated the moisture content, very substantially increased the density and bulk density and converted the material to fluid (as in the case of oil and gas) or a readily handled solid, as in the case of coal.

The question of thermal content per unit weight deserves special consideration. Although, theoretically, a fuel can be any chemical substance that reacts with another generating heat, fuels as commonly understood all involve carbon or hydrogen, or most commonly a combination of carbon with hydrogen, reacting with oxygen to produce carbon dioxide, water or both. In this context elemental carbon and elemental hydrogen can be thought of as the ultimate fuels; they themselves differ quite significantly in thermal content, the heat of combustion of elemental carbon being 34.4 GJ/t and the heat of combustion of hydrogen being 141.9 GJ/t. Elementary carbon is the principal component of coal and its heat of combustion equates quite closely to the heat of combustion of coal. Hydrogen, on the other hand, is not commonly available as a fuel, though if it were it would be highly suitable for certain uses, such as fueling aircraft, where the ratio of thermal value to weight is of crucial importance. Oil and gas normally fall between these limits, since they contain both carbon and hydrogen. For example, in the case of paraffins, the longer the carbon chain which makes up the molecule, the more nearly the proportion of hydrogen to carbon by weight approaches 1:8, corresponding to 12.5% hydrogen, the theoretical value for a chain of methylene units (CH_2) of infinite length. For example, in the case of n-octane the percentage of hydrogen is 15.8. Inevitably, among the hydrocarbons, the rules of valency prescribe that the carbon, with its atomic weight of 12, will always greatly exceed hydrogen, with its atomic weight of 1. Indeed, the greatest ratio of hydrogen to carbon occurs in methane, CH_4, in which hydrogen accounts for approximately 25% of the weight of the whole molecule. This gives methane a special position as the hydrocarbon fuel with the highest thermal content, 55.7 GJ/t compared with 44.0 GJ/t for a typical oil.

By contrast, the thermal content for ash-free biomass is usually close to 20 GJ/t once it is fully dried. Biomass containing significant quantities of moisture gives effective values much lower than this. The basic chemical reason for the lower heating value even of dry biomass, is its content of oxygen, and, to the extent that the biomass contains oxygen in combination with its carbon and hydrogen components, the material can be thought of as already partially oxidised or 'burned', making part of the carbon and hydrogen fuel unavailable. Table II, which illustrates the formulae of some common organic chemical

Table 1 Heats of combustion of major biomass components, major biomass types and major fossil fuels

Substance	Heat of combustion	
	kcal/kg	MJ/kg
Biomass components		
Cellulose	4500	18.8
Starch	4500	18.8
Fats	9300	38.9
Protein	5600	23.4
Lignin	6100	25.5
Biomass types (bone dry)		
Grass	4400	18.5
Wheat straw	4200	17.6
Peanuts, rapeseed	7050	29.5
Wood	4200	17.6
Meat	5800	24.3
Fossil fuels		
Oil	10 500	44.0
Bituminous coal	8000	33.5
Anthracite	8700	36.4

substances, and lists their heating value, illustrates clearly the progressive effect of the introduction of more oxygen into the molecule in reducing the thermal value. Substances such as methyl alcohol, ethyl alcohol, having only a single oxygen atom introduced, are still relatively good fuels with substantial thermal content. More highly oxidised compounds such as formic acid, glycolic acid or, worst of all, oxalic acid, have very modest thermal value. This concept can easily be applied to biological materials. These inevitably, as they arise in nature, comprise extremely complex mixtures of chemical components, some of which have higher heating values than others. This is illustrated in Table III which lists some major classes of biological substances according to their thermal values. The fatty acids, and the fats which contain them, have some of the highest thermal values to be found among biological material, and this can be ascribed to their relatively long hydrocarbon chain attached to only a single carboxyl group containing the two oxygen atoms. A fatty oil, therefore, is the closest biological equivalent to a fuel oil; fatty biological materials therefore have a much higher thermal content than non-fatty ones and biological processes to produce fats as fuel have some potential long-term interest for biofuel production. Much the same observation applies to waxes, whose main component is long-chain paraffin alcohols, and to substances formed by condensation of long-chain fatty acids, such as cutin (from the surface of leaves) and suberin (from bark). Other biological materials with higher than average thermal content are those whose structure is based largely upon benzene rings, such as lignin, a principal structural component of wood, which is based upon condensed phenylpropane subunits, and steroids, such as cholesterol, which contain a polycyclic aromatic nucleus. Lignin has a thermal content of 25.5 GJ/t. The biosynthetic systems of plants also produce some actual hydrocarbons, especially isoprenoid hydrocarbons (i.e. based upon isoprene subunits) as in terpenes, carotenes and in rubber. The chemical structures of these substances are given in Table III. Apart from special components of this kind, the majority of biomass material is at an oxidation level close to that of carbohydrate, which can be typically illustrated by the formula of glucose $C_6H_{12}O_6$, in which the hydrogen and oxygen occur in the same ratio as in water, so that the hydrogen may be thought of as fully oxidised, leaving in effect only the carbon that can still be burned as fuel. This is an oversimplified approach, because the actual heats of combustion of such compounds are influenced not only by the thermal value of their elemental constituents but also to some extent by the heat of formation of the compound concerned. Nonetheless, it is a useful concept to think of the hydrogen already being fully oxidised and the carbon available as fuel. The thermal value of glucose amounts to approximately 16 GJ/t, which rises to approximately 17.7 GJ per tonne for starch and cellulose, which have been formed from glucose by abstraction of water. Indeed, the thermal value of glucose per gram molecule of the carbon which it contains (0.48 MJ), is quite similar to the heat of combustion of a gram

Table II Heats of combustion of organic compounds at different levels of oxidation

Compound	Formula	Approximate heat of combustion derived from elementary composition (kcal per 12 g carbon)	Experimental heat of combustion (kcal per 12 g carbon)
1 carbon compounds			
Methane	CH_4	231.1	212.7
Methyl alcohol	CH_3OH	183.9	173.6
Formaldehyde	$H_2C=O$	115.5	136.4 (as a gas)
Formic acid	$HCOOH$	78.4	
Elementary carbon	C	98.4	98.4
Carbon monoxide	CO	49.2	67.6
Carbon dioxide	CO_2	Zero	Zero
2 carbon compounds			
Ethane	CH_3-CH_3	197.0	184.2
Ethyl alcohol	CH_3-CH_2-OH	173.3	163.8
Acetaldehyde	CH_3-CHO	139.2	139.4

FUEL USE AND CONVERSION TECHNOLOGIES

Table II (continued)

Compound	Formula	Approximate heat of combustion derived from elementary composition (kcal per 12 g carbon)	Experimental heat of combustion (kcal per 12 g carbon)
Acetic acid	H–C(H)(H)–C(=O)–O–H	115.6	104.5
Ethylene glycol	H–C(H)(OH)–C(H)(OH)–H	149.7	141.0
Glyoxal	O=C(H)–C(H)=O	81.4	
Glycolic acid	H–C(H)(OH)–C(=O)–O–H	92.0	83.1
Glyoxylic acid	O=C(H)–C(=O)–O–H	57.8	
Oxalic acid	H–O–C(=O)–C(=O)–O–H	34.2	29.4

molecule of elemental carbon (0.41 MJ). A very high percentage of most plant biomass is carbohydrate in the form of cellulose or starch and the high oxygen content of these carbohydrates accounts for the modest thermal value of most biomass.

The advantages of biomass fuels, such as they are, are not connected with suitability in storage and use. One advantage is that they constitute a continually renewable resource whose use leads to no long-term increase in the atmospheric carbon dioxide. Sometimes they may be cheap and readily available. When this coincides with the material being available in a sun dried condition, as it does in gathering firewood in a forest, when a farmer uses straw from his field to heat the farmhouse or when tea plantation workers use bush prunings that have dropped on the ground, then these may, indeed, be real advantages.

Table III Chemical structures of biomass substances

SUBSTANCE			
Fats (triglycerides of fatty acids)	$\begin{array}{l}CH_2OCOR_1\\CHOCOR_2\\CH_2OCOR_3\end{array}$	(where R = similar or dissimilar hydrocarbon residues)	
Fatty acids (examples)	stearic	$CH_3(CH_2)_{16}COOH$	
	oleic	$CH_3(CH_2)_7CH=CH(CH_2)_7COOH$	
	linoleic	$CH_3(CH_2)_4CH=CHCH_2CH=CH(CH_2)_7COOH$	
Waxes (examples)	cetyl alcohol	$CH_3(CH_2)_{14}CH_2OH$	(occur in combination with fatty acids)
	cocceryl alcohol	$CH_3(CH_2)_{18}CO(CH_2)_{13}CH_2OH$	
	n-triacontanol	$CH_3(CH_2)_{28}CH_2OH$	
Lignin (complete structure not characterised)	[structure of lignin fragment with CHOH, CHCH$_2$OH, CH$_3$O–, HOCH$_2$CH, CH–O groups on aromatic rings]		(small section only of polymeric molecule)
Terpenes and terpenoids (examples)	limonene	β-terpineol	
Carotene (a monocyclic diterpene)	β-carotene [conjugated polyene structure with terminal cyclohexene rings]		
Rubber	$---CH_2\underset{|}{C}=CHCH_2CH_2\underset{|}{C}=CHCH_2CH_2\underset{|}{C}=CHCH_2---$ with CH_3 groups (part of large polymeric structure based upon isoprene)		

FUEL USE AND CONVERSION TECHNOLOGIES

The need for biomass as fuel runs directly counter to the biological need of the soil to have organic matter added to it. Compromise will often be necessary between these two requirements. However, it is not only the carbon component of the biomass which may be important to future soil fertility, but also the ash components, containing potassium, phosphate, trace elements and nitrogen. Nitrogen is often of especially crucial importance in the productivity of biological systems and, if it is not recycled, it must be replaced either by increased biological nitrogen fixation or by artificial nitrogen fertilisers. These factors can have a considerable economic impact upon the prospects for certain biomass energy schemes, and, even if not dictated by economics, must always be borne in mind in respect of their long-term environmental impact. In particular, the combustion of biomass or its treatment by thermal conversion processes followed by subsequent combustion, destroy the organic nitrogen and leave only ash. Only biological conversion processes have the potential to conserve the nitrogenous component of biomass for recycling to the soil; this is potentially important on account of the high energy cost of manufacturing replacement nitrogen fertiliser. Biomass containing 12% of crude protein in dry matter contains approximately 19 kg of N/t and its replacement with synthetic nitrogen fertiliser would consume over 1 GJ/t of biomass; this is therefore an item of the debit side of the energy balance whenever biomass is either combusted directly or thermally converted without ammonia recovery.

A1 PURPOSES OF CONVERSION PROCESSES

The concept of a conversion process is not restricted to biomass. Crude oil has to be refined to make it suitable for domestic heating or for use in automobiles. Coal may have to be sieved and graded for use in certain types of combustion apparatus or to make it compatible with handling equipment. A more radical degree of processing occurs when coal is processed and converted into town gas and coke, when coal is hydrogenated to produce an oil or when coke is gasified to make producer gas or water gas. Indeed, electricity generation using coal, gas or nuclear fuel is really another form of conversion process not involving biomass. The usual reason for introducing a conversion process with fossil fuels is to adapt the fossil fuel to a particular end use. In the case of biomass there are additional reasons, represented by the task of overcoming the various disadvantages of biomass as a fuel that have been set out above. Removal of moisture is often a major objective, which may involve consumption of some of the fuel in dehydrating the remainder or, alternatively, when the finished fuel product is a gas or volatile liquid, the converted fuel may be separated from the moisture by volatilisation. The removal of moisture necessarily fulfils another objective — that of preserving the fuel.

To simultaneously increase the thermal content and decrease the bulk density is a most important objective. Since the modest thermal content is brought about by the oxygen component in the biomass, increased thermal content necessitates either driving off oxygen in the form of water, or volatilising the fuel component such as ethanol or methane. Loss of water may arise either from thermal decomposition of compounds in the organic material, or, alternatively, from enzymically induced rearrangements whereby some molecules become reduced and others oxidised. Such rearrangements are at the basis of all biological conversions, some of the carbon being reduced to ethanol or methane while much of the remainder is oxidised to carbon dioxide; the biological evolution of elementary hydrogen from biomass is a special case, but it nonetheless must leave the unvolatilised carbon constituents at a higher level of oxidation than in the feedstock.

Sometimes all that is required is to change the handling characteristics of the fuel, for example, to make it free flowing, as when straw is finely chopped or ground or when municipal waste is converted to a granular form. The production of a fluid fuel is often an important objective, which may be either gas or liquid, since this implies that it can then be pumped and metered by automatic equipment. The special requirements of the motor car engine and the inconvenience of carrying a gas on a vehicle, have tended to dictate that ethyl alcohol, or methyl alcohol have been almost the only forms of biomass-derived fuels considered for car engines.

Finally, another objective of conversion, though it is usually a subsidiary one, may be that of separating the desired carbon and hydrogen elements of the biomass from the remaining components which may be valuable in another context, such as animal feed or fertiliser.

A price has to be paid for accomplishing these objectives. Any processing plant involves capital costs, materials and labour; it inevitably involves the consumption of some fuel in the processing operation itself and wastage of some of that fuel. The price to be paid is therefore measured in financial terms by an increase in the cost of the finished fuel, of which one component comprises the various non-fuel inputs to the process and the other comprises the loss or consumption of a proportion of the feedstock. A concept develops, therefore, of what is meant by energy conversion efficiency. In the simplest terms this is given by the thermal value of the output divided by the thermal value of the input, a formula which takes into account any of the original fuel used up or lost in the process itself. However, this is not really sufficient, since one must also make allowance for any additional energy used in the process from outside sources such as electricity from the mains, the energy content of any chemicals used, such as sulphuric acid or purified oxygen, the energy cost of manufacturing the plant for the process and, finally, the energy cost of harvesting the biomass and transporting it to the plant. From this one can develop an overall energy budget showing how worthwhile it is, in energy terms, to conduct the process.

FUEL USE AND CONVERSION TECHNOLOGIES

B Types of conversion process

There are actually only two principal classes of conversion process, thermal and biological. These convert biomass into useful fuels via heat treatment and microbiological action, respectively. However, there are more than just these two types of basic process to be considered because biomass may be utilised by direct combustion, i.e. without conversion, and this may raise different engineering and technological problems from the direct combustion of fossil fuels. Either direct combustion or conversion processes may require that the biomass should first be subjected to preliminary processing; moreover the fuel first produced by thermal or biological processes may require secondary processing to upgrade their quality or change their physical state by chemical modifications as when methane gas is converted to methanol as an automobile fuel. Preliminary processing, prior to direct combustion or thermal or biological conversion processes, may involve dehydration, comminution and other alterations of physical form. Gas produced from biomass by conversion processes may require cleaning to remove contaminants, especially carbon dioxide, or they may be subjected to a shift reaction in which one component of gas becomes more oxidised and another more reduced followed by separation of the two components, in order to give a gas of increased thermal content.

A third class of conversion process may eventually need to be added under the heading 'aqueous reductive', since it has been shown (Bliss and Blake, 1977) that an alkaline slurry of cellulose at 250–400 °C under pressure can react with carbon monoxide to give a substantial yield of a low-oxygen oil with a thermal content only 10% below that of petroleum oil. It appears that carbon monoxide abstracts oxygen and undergoes conversion to carbon dioxide. This process is a recent conception and needs to be more widely studied. If it establishes a place among conversion technologies, it could be regarded as thermal, though it is clearly distinguished from thermal processing as normally understood by its occurrence in an aqueous medium.

C Direct combustion

The direct combustion of biomass to produce heat should in theory be an efficient route of utilisation, since, with no conversion process involved, there are no conversion losses. On the other hand, the advantage from conversion processes — such as moisture reduction, ease of handling and increase of thermal value, are not available and the boiler or furnace designer must grapple with these problems. The almond shells often used as boiler fuel in Mediterranean countries are as near perfect for fuel use as any unmodified biomass material is likely to be, having already been baked dry in the sun, having a useable density and being naturally formed in regular size pieces that facilitate automatic stoking.

This can be contrasted with attempts to burn wet leaves or garden weeds. Biomass rarely arises naturally in an acceptable form for burning; straw is of low bulk density until compressed and requires chopping or grinding for automatic feeding, either in the automatic stoker or beforehand and, although wood may be burnt as logs, such large pieces generally combust too slowly to be efficient in industrial equipment, boilers generally requiring chipped wood. These preliminary processes of chipping, chopping, grinding, densification or drying are associated with financial costs and energy expenditure. Where these problems have not been fully resolved, the combustion process becomes inefficient by wasting heat in volatilisation of moisture, heat losses from the equipment and by demanding excessively large and expensive equipment in relation to heat yield.

Moisture is a major problem in countries with predominantly damp climates where, for most of the year it is impossible to produce anything other than wet biomass and where the weather precludes natural drying for most of the year. Such material may well be better utilised as feedstock to anaerobic digestion, where the process of gas formation is effective in largely separating fuel from water. It is interesting, however, to consider the likely limits of direct combustion using moist biomass, especially since it is commonly stated that material for combustion as fuel should not be above 15% with respect to moisture. In practice this is far from the truth because, to achieve combustion, all that is required is that the biomass should contain sufficient gross thermal value to volatilise the water that is present, to raise it to the temperature of the flue gases and to keep the combustible material sufficiently hot to enable combustion to continue. Although the latent heat of evaporation of the water is often thought of as the major obstacle, and it is, indeed, significant in the calculations, its effect is not as serious as is often supposed. The latent heat of evaporation of water amounts to 2.4 MJ/kg; hence a kilogram of dry solids, having a typical thermal value of 17 MJ, will, during its combustion at 50% moisture content, lose some 14% of gross thermal value in volatilising its water content; if the exit gases were at, say, 700 °C, an equal quantity of heat would be used in heating the moisture to this temperature; heat losses in flue gases and through the boiler wall have still to be allowed for, but it seems likely that a 60% boiler efficiency may be achievable. On a comparable basis, biomass at 65% moisture content would be expected to give an efficiency of only about 25% and, inevitably, at that level, special measures would become important, either to dry the fuel at low temperature prior to combustion, or to recover the latent of steam and/or exhaust gas heat. Sugar cane bagasse is widely burnt at sugar factories to provide energy for the process and for this purpose it is usually pressed to about 50% moisture. Even wetter materials can be usefully burned, as is shown by the work of Potter and Keogh (1979) with high moisture content brown coals; at 70% moisture these gave a maximum boiler efficiency of 60%, which rose to 67% at 60% moisture;

a practically attainable efficiency in each case is likely to be of the order of 48% and 54% respectively. Brown coal usually has a thermal value of about 28 GJ/t and will therefore continue to give sustainable combustion at high moisture contents which could render typical biomass incombustible unless outside heat is applied; a better example in many respects is the application of fluidised bed technology to incineration of sulphite liquor at 35% solids, giving self-sustained combustion, in a process worked out by the Columbus Laboratories of the Batelle Memorial Institute that resulted in commercial application at the Carthage, Indiana mill of the Container Corporation of America (Smithson, 1977). These examples lead one to expect further activity in the same area of technology, whereby wet biomass will be used, accepting a degree of thermal inefficiency during combustion or prior drying as an alternative to incurring process losses and, possibly, even higher capital expenditures, for processing into liquid or gaseous fuels.

The densification of biomass is often a priority as well as reduction in moisture content. Thus, straw may be compressed into dense bales (see chapter 4) and can be further compressed into quite dense blocks or briquettes if some of the fuel value is sacrificed in a preliminary partial pyrolysis step (Hansford, 1974); densities as high as 1.0 have been claimed, although the result, when blocks are made from bales, are inevitably quite large pieces of fuel that would burn rather slowly in industrial equipment. Conversion of straw into dense chips or granules that would be flowable and ideal as boiler fuel does not appear to have been accomplished, and it is possible that the economic cost of such modification would always be too high.

The chief problem with wood as fuel is not so much density (although biomass is almost always less dense than solid fossil fuels) and moisture content, as size reduction. The slow combustion of saw-logs may be acceptable domestically but not industrially, and the toughness of timber results in substantial energy costs for chipping. Straw chopping is a similar but less energy consumptive process, enabling the fuel to be moved in and out of storage and into the furnace by an air current.

The choice between direct combustion and one or more of the conversion processes is therefore to be made based upon a combination of the fuel moisture content, density, thermal value and physical form, especially as related to mechanical handling. Ash content is also a parameter of some significance in comparison to ashless fuels such as oil and gas.

Once direct combustion has been decided upon, the choice of boiler or furnance is related to these same variables, though it is also dependent upon the scale of operation and the end use of the energy. Simple *water-jacketted boilers* are suited to domestic use, as with other fuels, though they nonetheless have to be increased in size to accommodate sufficient straw or wood to give the same heat output as has been obtained from fossil fuels.

When large-scale production of high pressure superheated steam is required for power generation, a *water tube boiler* is the commonest choice, and the same choice generally applies to the generation of high to medium pressure steam for industrial processes. This type of boiler has a large combustion area surrounded by banks of vertical water tubes; an important feature of biomass fuels is their high content of volatile matter compared to solid fossil fuels; the large combustion chamber required for biomass may be dictated, not only by the lower density and bulk density of the fuel, but also by the need for ample space above the fuel bed to prevent flaring volatile material to impinge upon the chamber wall and, in time, damaging it. Because it embodies a large combustion chamber, the design of the water tube boiler gives it some flexibility as to fuel type; compared with the shell boiler (see below) uniform and high grade fuel is not required and final steam temperatures may be up to 500 °C. Such boilers have quite frequently been installed for combusting sugar cane bagasse, wood waste and domestic refuse; hence, there is no doubt about the feasibility of such operations though, with biomass fuel, the industry has still to optimise boiler design, efficiency, capital cost and stoking methods. A shell boiler consists of a furnace tube, which in many designs may be quite narrow surrounded by an inner and an outer bank of tubes, the hot gasses passing from the furnace tube and successively over the inner and outer banks of tubes. Such boilers are usually applied to raising relatively low pressure steam for industrial processes and heating. These are generally rather unsuited to the combustion of biomass fuels on account of the restricted diameter of the furnace tube and the susceptibility of the tube wall to damage by flame impingement, which is a particular problem with biomass. Moreover, they are relatively demanding as to fuel quality and state of division.

Other forms of combustion equipment best suited to size-reduced biomass include *cyclone furnaces* and *fluidised bed systems*. In the former, the combustion chamber is cylindrical and high velocity air is admitted tangentially to create a vortex within the equipment; fuel may be injected in finely divided form, intended to remain in suspension while burning, or placed on a hearth at the base.

In fluidised bed systems fuel is introduced into a bed of sand, or similar inert finely divided material, fluidised by an air-stream. Either of these systems may be used as a source of direct heat, for example for drying, or used to generate steam via a waste heat boiler. Many different versions of these systems have been devised, especially fluidised bed systems, and, as they are both fairly recent introductions, designs have not necessarily been optimised. The fluidised bed systems have not been widely applied to biomass fuels and difficulties arise from their lightness, which predisposes them to be blown out of the bed, and their high volatile content, leading, as before, to flaring above the bed. On the other hand, cyclone furnaces have proved their adaptability to biomass, working on inputs such as wood waste and reject tyres. An interesting development in the combustion field that is of special relevance to biomass waste is the *controlled combustion incinerator*; the interest arises from its being well suited to combusting heterogeneous organic materials, such as may be derived from municipal solid waste, and because it bridges the two processes of direct combustion and thermal conversion. In a controlled combustion incinerator there are two combustion chambers, the first one being large and supplied with only a small proportion of the air needed for complete combustion of the fuel. A gasification process necessarily ensues and the gas passes to the second chamber where it is admixed with ample air supply and supplementary fossil fuel; it is therefore similar to a combined gasifier and gas boiler, although the need for supplementary fuel must be regarded as a disadvantage.

In all systems ash removal must be given consideration; procedures exist to segregate ash particles from the bed material of fluidised bed systems and various types of grate are available, accommodating larger pieces of fuel, from which the ash may be mechanically removed.

In solid fuel systems not employing size-reduction and suspension, the mechanical stoking is onto a grate, designed to accommodate the particular fuel.

The grate may be stationary or rotating, or the fuel may be carried down the length of the furnace by chain grate stokers, and many different alternative means exist for mechanical stoking mainly developed for coal; in particular, the feed may be from above, from below, or, in effect from the side in the case where fuel is ram fed intermittently.

Boilers or furnaces developed especially for biomass combustion are generally variants of one of these designs, modified, especially with regard to dimensions per unit of heat output and fuel handling. Wood waste, domestic refuse and bagasse are perhaps the commonest fuels at present, but systems are under development to combust straw on scales above domestic size, either in chopped form or complete rectangular or large round bales (Strehler, 1979).

D Thermal processing

D1 THE NATURE, OBJECTIVES AND ADVANTAGES OF THERMAL PROCESSING

It has already been noted in section A that fossil fuels, which originated from biomass, have been put through a geological conversion process which overcame the principal disadvantages of biomass as a fuel. Conversion of biomass to coal involved a biochemical stage of transformation in which a peat was formed under largely anaerobic conditions, followed by a geochemical stage in which the peat was subjected to the pressure of superimposed layers of sediment or rock and to an increase in temperature, the type of coal being determined by the nature of the original biomass, the degree of anaerobicity of its decay, the temperature and pressures reached during the geochemical stage and the age of the seam. During these processes the carbon content has been increased, the hydrogen and oxygen contents diminished, the calorific value increased and the content of volatile matter diminished (Francis, 1961). Although a broadly similar pattern of events probably gave rise to petroleum deposits, and natural gas, the differences, which led to much more hydrogen remaining, are not clearly understood. Attempts to reproduce these processes in the laboratory, let alone in industrial processes, have been largely unsuccessful, but thermal processing of biomass may be seen as an attempt to substitute for the natural processes of conversion. Thermal processing has been quite widely applied to coal, i.e. to an already converted fossil fuel, and when this is done the objective is normally a change of state to yield a liquid or gaseous fuel.

In a now obsolete process for production of methanol, acetic acid and acetone, wood was heated in the absence of air to a temperature of about 250 °C. The feedstock decomposes at this temperature yielding (*i*) wood gas, containing hydrogen, methane, carbon dioxide and some minor components, (*ii*) pyroligneous acid, which is an aqueous distillate containing the methanol, acetic acid and acetone, (*iii*) a wood tar, rich in phenolic substances and (*iv*) charcoal, consisting of carbon with the original mineral or ash component of the wood. The yield of the desired components was only 5–6% based upon the original weight of wood. However, the process used to provide fuel incidentally to its main objective in the form of the gas, which was usually used to heat the retorts, and charcoal as a combustible solid fuel. This constitutes a simple form of thermal processing, although the treatment temperature was very low and the range of products excessively diverse. Processes of this type, conducted in the absence of air and with no other additions to the system, are known as pyrolysis (i.e. splitting by heat); the nature of the products is markedly affected by the temperature employed and by the rate of heating the feedstock. In Australian work, for example, carried out at 500–650 °C (McCann and Saddler, 1976) a yield of a pyrolytic oil from straw was obtained, amounting to some 39% of the input energy in the feedstock, although this was of poor quality, containing a significant amount of oxygen and hence low in thermal content per unit weight, compared to petroleum, and the process was not considered to be economic. Nonetheless, this process serves to illustrate the principle and aims of the thermal processing, which always involve converting the inconvenient, low calorific value and sometimes moist biomass fuel into a dry convenient fuel of increased calorific value, either an oil or a gas; advantage applies particularly to process versions in which the feedstock is converted with high efficiency to one form of fuel, either oil or gas, because a multiplicity of products complicates the process, increases investment for a given annual value of output and complicates marketing and distribution. In practice this usually necessitates one or more of the following steps; (*i*) optimisation of operating parameters to maximise the

FUEL USE AND CONVERSION TECHNOLOGIES

percentage of the desired fuel in the primary reaction products, (ii) introduction of additional reactants into the primary reaction chamber, especially, air, oxygen, steam or hydrogen and (iii) secondary processes to convert, modify or upgrade some or all of the primary products.

The advantages of the resulting fuels compared to the feedstocks relate to much greater ease of transport and distribution, greater ability to meter in a controlled way into combustion chambers via automated equipment, increased effectiveness in electricity generation, lack of ash and suitability for fuelling transport. To secure the full benefits of these various advantages, quite considerable modifications to primary reaction products are often necessary.

D2 DIFFERENT VERSIONS OF THE PROCESS AND CONDITIONS OF REACTION

(a) *Carbonisation or pyrolysis*

This is the basic version of the process in which biomass is simply heated in the absence of any air or additional reactants of any kind. An amount of char is always formed. Carbonisation is the name given to the process when the amount of char is maximised, the basic reaction being a driving off of water from typical biomass at the oxidation level of carbohydrate, i.e.

$$C_{6n}(H_2O)_{5n} \rightarrow 6nC + 5nH_2O$$

In practice, this reaction cannot be conducted with high efficiency, being complicated by other reactions, one of the most important being considered to be:

$$C + H_2O = CO + H_2$$

which is responsible for generating what are probably the main primary constituents of pyrolysis gas and the main components which are obtained in the upper range of temperatures. Several secondary reactions occur, such as:

$$2CO + 2H_2 \rightleftharpoons CH_4 + CO_2$$

$$C + 2H_2 \rightleftharpoons CH_4$$

$$C + 2H_2O \rightleftharpoons CO_2 + 2H_2$$

so that methane and carbon dioxide are also present in various proportions in the pyrolysis gases, depending upon conditions, upon the feedstock employed and its moisture content. A wide range of gas compositions have been reported in the literature from various feedstocks, the interesting changes of gas composition taking place between 500–1000 °C. Experimental work with municipal waste and with wood confirms that the volume percent of hydrogen and carbon monoxide in the gas increases steadily above 400 °C until they become almost the sole gaseous products at temperatures approaching 1000 °C, (Lewis and Ablow, 1976). High yield of gas is also favoured by high temperature with yields of oil, char and aqueous distillate depressed and Douglas *et al.* (1974) found that with rapid heating and with paper as a feedstock a gas yield on a w/w basis of over 50% of the charge was obtainable at 900 °C, but less than 10% at 400 °C, at which temperature about 35% of the product was char. Work on heating finely divided biomass very rapidly suggests that even higher overall gas yields are obtainable. Knight (1976), using a sawdust feedstock, found the yield of oil to remain fairly constant, at around 16–17% of the feed, over the range 540–870 °C. The rate of heating, however, appears to be quite significant in relation to the ratio between char and liquid products, slower heating tending to maximise the formation of char. In view of the premium values place upon liquid fuels, especially for powering automobiles, it may well be that the main future role for simple pyrolysis procedures will be in the production of liquid fuels, gas being more efficiently obtained in processes involving injection of air or oxygen. However, the poor properties of primary pyrolytic oil, which have already been referred to in section D1, are exacerbated by corrosiveness, high ash and a tendancy to polymerise. Considerable modification after pyrolysis would be needed to make this an acceptable fuel for the market. For example, one approach is to subject the oil to a 'gas phase pyrolysis' step or vapour cracking to yield olefins as precursors to alcohol or gasoline.

(b) *Air or oxygen gasification*
When biomass is heated in the presence of some air or oxygen, but insufficient to combust it, gas formation is generally maximised and temperatures within the reactor rise on account of the oxygen-consuming reactions, especially:

$$C + O_2 \rightarrow CO_2$$
and
$$C + \tfrac{1}{2}O_2 \rightarrow CO$$
and
$$CO + \tfrac{1}{2}O_2 \rightarrow CO_2$$
and
$$CO_2 + C \rightleftharpoons 2CO$$

though methane and hydrogen formed simultaneously by the thermal splitting of the organic material may also be combusted; carbon dioxide formed may also be reduced by hydrogen present in the gaseous mix:

$$CO_2 + 4H_2 \rightleftharpoons CH_4 + 2H_2O$$

The resulting gas tends, of course, to be higher in carbon dioxide than in the case of pyrolysis. Both pyrolysis gas and gas obtained by oxygen gasification tend to be in the medium energy range, having a thermal content of 11–19 MJ/m^3. However, when air is used as the oxygen source for the reaction about half of the product gas is likely to be nitrogen, passing through the process unchanged, and decreasing the thermal value of the gas to 4–8 MJ/m^3, a disadvantage which renders it relatively unsuited to subsequent upgrading to synthetic natural gas (SNG) or to methanol. A major advantage of using an oxidative process with air or oxygen is that the char normally remaining after pyrolysis is removed by gasification, thus maximising the overall output of gas from the process. Indeed, it is an option that is open following a pyrolysis process to gasify the chars, tars and oils by air or oxygen gasification. However, air gasifiers operating on whole biomass feedstocks may be so designed either to permit or to eliminate the formation of oils and tars, or at least to allow or prevent their appearance in the product gas. Avoidance of oil and tar is by so designing the equipment to prevent their escape except via the hottest region of the bed in which they are destroyed. With simple gasifiers that allow oil and tar to distill over with the gas, it seems best to burn the gas on site before they condense, otherwise a rigorous cleaning operation would be needed. Gas from air gasification is often referred to as 'producer gas', a name first applied to gas obtained from coke by a similar process.

The admission of oxygen or air fulfils the criterion already noted for thermal processing that one type of fuel product should be maximised, to the point if possible of becoming virtually the sole product. Maximising of gas yield is fairly readily achieved by thermal processing, whereas no one-step process to yield a predominantly liquid product of marketable quality has yet been demonstrated.

The composition of the gas obtained by oxygen gasification is broadly similar to pyrolysis gas obtained at high temperature of reaction, in that hydrogen and carbon monoxide predominate, but there is an inevitable increase in carbon dioxide and water vapour. For example, in work on the oxygen gasification of manure at 29% moisture, a Stanford Research Institute team led by Alich and Inman (1976) found the gas to comprise 43% CO, 26% H_2, 12% CO_2, 9% H_2O and 4% CH_4. The temperature of the reaction was 1650 °C, sufficient to melt the ash to a slag, but exit gases were only at 105 °C. Typical values for air gasification are similar to those quoted by Lucas (1979) for air gasification of wood as 16% CO, 20% H_2, 50% N_2, 12% CO_2 and 2% CH_4, with a reaction temperature around 1100 °C and a gas exit temperature of 600 °C. A most crucial parameter in determining product composition and operating temperature in air or oxygen gasification is the equivalence ratio, i.e. the ratio between the amount of air or oxygen admitted to the reaction chamber and the amount that would be required to completely oxidise the feedstock. It has been shown that the total energy content of the product gases is maximum when this ratio has a value of 0.25 or 0.3; in this same region the char fraction disappears as air or oxygen is admitted, increasing the eqivalence ratio from zero (i.e. pyrolytic conditions) to 0.25; at the same time carbon monoxide content reaches a peak at this equivalence ratio, while methane diminishes to very low levels. Increasing equivalence ratio similarly increases operating temperature as increasing combustion reactions release more of the feedstock energy. However, in the preferred range of 0.25–0.3 the temperature is typically 1000–1200 °C for an air gasifier and 1300–1750 °C for an oxygen gasifier. These temperatures govern the final forms of the ash, which is usually powder below 1100 °C and slag at more than 1300 °C. The rates of reaction, especially the rate of gasification of the first-formed char, are governed to a significant degree by particle size and also by the porosity of these chars, which influence their surface accessibility for reactants.

(c) *Gasification processes involving the use of water or steam*
When steam is blown through a bed of hot carbon, for example, incandescent coke at about 1000 °C, a reaction takes place to yield gaseous products, i.e.

$$C + H_2O \rightleftharpoons CO + H_2$$

and at this temperature roughly equimolecular proportions of carbon monoxide and hydrogen are obtained, and the product is known as 'water gas'. In the case of biomass gasification, water or steam may be added, but more commonly moisture is present in the feedstock. In one sense the presence of moisture is a burden on the energy balance of the process because it must either be vaporised or reacted, and its reaction with carbon is endothermic; on the other hand this reaction with steam is an alternative way of consuming char, which is normally an undesirable end-product, and maximising gas output. When steam reacts with an already formed char, the reaction is essentially the same as the steam gasification of coke, but fresh biomass at the oxidation level of carbohydrate requires very little additional moisture for maximum conversion to carbon monoxide and hydrogen by:

$$C_{6n}(H_2O)_{5n} + nH_2O \rightarrow 6nCO + 6nH_2$$

Therefore, it is usually found that moisture contents much above 20% by weight of the feedstock tend to be disadvantageous to the overall energy balance. Since the steam reaction is endothermic, it is best to admit some air as well to keep the process autothermic.

(d) *Thermal processes using added hydrogen*
Since one of the chief disadvantages of biomass as a fuel is its modest thermal content per unit weight, any process intended to upgrade it in this respect must either aim to separate some relatively oxidised components from the relatively reduced components which then remain as an improved fuel, as, for example, when CO_2 is removed from a CO_2/CH_4 mixture that has been derived from $(CH_2O)_n$, or, if the whole biomass carbon is to be utilised as fuel, then reducing power must be added from an outside source. The objective can be either to maximise the yield of methane by:

$$C_{6n}(H_2O)_{5n} + 12nH_2 \rightarrow 6nCH_4 + 5nH_2O$$

or to optimise the production of paraffins by:

$$C_{6n}(H_2O)_{5n} + 6nH_2 \rightarrow 6n(CH_2) + 5nH_2O$$

FUEL USE AND CONVERSION TECHNOLOGIES

The development of such processes for biomass is in its early stages and is tending to concentrate on methane formation. As matters stand at present, dependence upon added hydrogen will generally necessitate dependence upon the use of fossil fuel.

D3 PROPERTIES OF BIOMASS RELEVANT TO THERMAL PROCESSING

Although biomass suffers from several disadvantages as a fuel, it possesses certain advantages in thermal processing compared with coal, which apply particularly when the objective is maximum gasification. It contains oxygen and water, which are necessary for converting the carbon into gaseous fuels, and has a high volatile content. Biomass also typically contains more hydrogen relative to carbon than in coal and has a lower content of sulphur, i.e.

	Coal (rich in vitrinite)	Typical biomass
Volatile matter (% of dry matter)	15–40	70–90
Oxygen (molar ratio to carbon)	0.02–0.1	0.4–0.9
Hydrogen (molar ratio to carbon)	0.55–0.75	1.5–2.0
Sulphur (% of dry matter)	2.0–4.0	0.1–0.5
Ash (% of dry matter)	5.0–25.0	5.0–25.0
Thermal content (MJ/kg)	32.0–35.0	17.0–22.0

Other properties that are relevant to the behaviour of biomass during gasification are its poor heat conductivity and the porosity of its char.

D4 PREPARATION OF BIOMASS FOR THERMAL PROCESSING

Biomass may require one or more steps from among densification, adjustment of particle size by pelletising or chipping, dehydration, removal of incombustible objects, before it is sent to thermal processing.

For the reasons already discussed, excessive moisture in the feed is undesirable. At moisture contents much above 65% on a w/w basis, moisture can usually be expressed with either a flat bed or screw press, the maximum solids content attainable depending upon the nature of the material; to get moisture down to 20% or better dehydration must be employed, although quite low-grade heat such as process exhaust gases may be used for this purpose.

Control of feedstock particle size is important because it affects reaction rate, because a bed of more or less standard size particles is likely to be more efficient than a heterogeneous bed of assorted sizes, and because the operation of some types of equipment demand a particular size range. Fixed bed gasifiers require discrete chunks of material of several centimetres diameter, whereas fluidised bed gasifiers require much smaller particles but again, uniformity of size aids the operation. Densification is important too because biomass is bulky relative to its weight, decreasing the maximum throughout of thermal processing equipment and increasing investment levels relative to a given output of converted products. Densification in the form of pellets may yield a form of pre-treated biomass having a density of 1.0 or more, overcoming this problem, and particle size is standardised within the same process.

Wood may be chipped to a 1 to 1.5 cm size for an energy expenditure as electricity of only 1% of the gross thermal value of the wood however, when the inefficiency of gasification and of electricity generation are both taken into account, this figure becomes almost 5%. Similarly, straw, or other small but fibrous biomass may be chopped. Either the chopping of straw or the chipping of timber, or indeed, the shredding of municipal waste, may be necessary prior to pelletisation. Since the energy consumption in size reduction is a significant factor, attempts have been made to reduce this by chemical treatment, especially when a fine state of division is wanted, as in the case of ECO-FUEL II, a process developed by Combustion Equipment Associates Inc., which uses hot ball-milling in the presence of a chemical 'embrittling agent', and quite a fine dust results, understood to be about 0.2 mm particle size, though there is a substantial energy penalty for heating the mill and its feed.

D5 TYPES OF THERMAL PROCESSING EQUIPMENT

The multiplicity of the different chemical versions of thermal processing is reflected in the wide range of equipment types which have been produced to contain the biomass (or coal), to move it through the system, to impart the necessary heat to the feedstock and to control the through flow of gases, with or without the option of recycling these.

Most industrial versions will necessarily have to be continuous, with uninterrupted input of feedstock and output of gas. Such systems may be divided between those which are inherently autothermic, i.e. generate their own self-sustaining heat internally and those which require added heat, in the form of either combusted reaction products or from bought-in fuels from sources unconnected with the process itself. When external heat has to be applied, it may be done by means of a burner directing flames or hot gases onto the outside wall of the reaction chamber, though this results in only relatively inefficient heat transfer and the rate of heating each freshly added increment of feedstock is quite slow. These problems may be overcome by passing externally heated gases into the reactor chamber, either combustion gases from a separate burner consuming char, process gas or an unrelated fuel, or recycled process gases that have been heated in the separate burner after exiting from the process reactor. Superheated steam may also be employed to apply externally generated heat, but suffers from the limitation that above a certain proportion the addition of further moisture is disadvantageous to the reaction chemistry and energy balance.

In some process versions the feed is mixed with a particulate material of relatively high specific heat which is inert under the reaction conditions, to increase the heating rate and hence, influencing the composition of the reaction products. This feature applies for example, to a fluidised bed reactor which would operate with quite small particles of biomass in fluidised sand, just as in a fluidised bed boiler or furnace (section C). However, ceramic or steel spheres may also be used or even moten salts or metals, to achieve close contact of a high specific heat solid with the feedstock (Crane *et al.*, 1975; Soeth, 1973).

The simplest shape and design of the reactor itself is the *fixed bed reactor*, which consists of an upright metal cylinder containing the bed of feedstock, having an inlet and outlet for gases, a means of feeding additional solid feedstock from above and a means of ash removal at the base. There are different versions of this basic concept depending upon whether the flow of gases is upward and counter-current with the flow of solid feed (*updraft type*) or downward and co-current with the movement of solid (*downdraft type*) or simply across the bed (*cross-flow type*). These three types and their mode of operation are illustrated in Fig. 1. The results, in terms of the composition of the products are different;

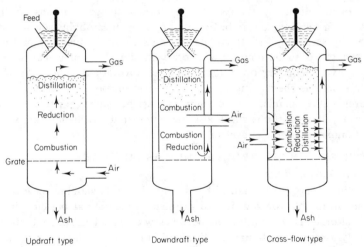

Fig. 1 Gasifier designs

the very simple updraft gasifier permits the vapours of liquid organic products and tars to distill over, while in the downdraft design, such materials cannot find an exit from the reactor without passing through the highest temperature zone, in which they are destroyed. The cross-flow type of reactor does not have this advantage to the same extent as the downdraft type, but both the cross-flow and downdraft designs serve to prevent the re-contracting of the reaction products with the freshest feedstock.

An important difference from these designs is the *rotary kiln*, in which the reactor is a rotating drum moved by motor-driven rollers and slightly inclined from the horizontal. The resulting tumbling of the feedstock assists heat transfer, particularly when heat is being applied through the reactor wall. There is also a family of reactor types where the feedstock is in suspension in gases, either with or without an inert heat-transfer material, of which perhaps the most important are the *fluidised bed reactors* and *cyclone reactors* based upon the same principles as fluidised bed and cyclone furnaces (section C). A further group of designs involves moving the feedstock on a flat reactor bed (e.g. *horizontal moving bed type*) or on a series of such beds positioned at different levels, with solid material moving progressively from one level to the next one down, while gases flow counter-currently and are vented from the top. In all these cases the hardware is usually of steel but, depending on the temperature reached, may be lined with refractory material.

D6 GAS CLEANING

Since gas appears to be the most economically viable primary product of thermal processing at the current stage of development, attention is now turned to the further treatment of the gas. As it is first formed the gas contains fine droplets of oils and tars and these may be removed by a water scrubbing stage or, in some versions, by scrubbing with the oil itself. This step can normally be omitted only when the gas is to be consumed at once on site. Scrubbing may have to be followed by a secondary operation such as electrostatic precipitation. When the gas is required as chemical feedstock or for conversion to methanol as a liquid fuel, even higher standards of cleanliness must be applied; in this case sulphur and certain hydrocarbons must also be eliminated, to avoid inactivating the catalyst employed in the conversion.

D7 UTILISATION OF PYROLYSIS OR GASIFIER GAS

(a) *Synthetic natural gas*

The medium energy gas obtained by pyrolysis is quite suitable for local captive uses but unsuitable for addition to gas mains which are at present piping high energy natural gas. If gasification becomes widespread in future there may be eventually a case for national gas distribution networks, short of natural gas, to change to the distribution of medium energy gas and thus to admit clean gasifier or pyrolysis gas without need for further processing, though this would incur a penalty in the form of reduced effective transfer within existing distribution systems. The alternative is to upgrade the primary, medium energy gas, into high energy gas, i.e. synthetic natural gas, (SNG), in which the principal product is methane. This is accomplished by the exothermic reaction:

$$CO + 3H_2 \rightarrow CH_4 + H_2O$$

The necessary increase in the hydrogen content of the gas mixture is obtained by the 'shift' reaction,

$$CO + H_2O \rightarrow CO_2 + H_2$$

and an iron or Raney nickel catalyst is employed at 400 °C. The catalyst is closely contacted with the catalyst in reactors of a wide range of possible designs, the catalyst usually being adherent to the surface of tubes or to particles of an inert support material. In such a process developed by Union Carbide and employed by Stanford Research Institute in methanating medium energy gas derived from manure, the following compositional change was noted (Alich and Inman, 1976) on a mol% basis:

	Feed gas	Methanated gas
CO	47.6	trace
CO_2	13.3	2.6
CH_4	3.8	68.5
H_2	28.2	6.1
Ar and N_2	6.0	19.6

quoting main constituents only. The energy content of the gas was increased from 11.6 MJ/m^3 to 29.7 MJ/m^3.

(b) *Methanol*

In view of the premium necessarily placed upon the value of liquid fuels, conversion of the primary gas output from thermal processing to methanol is one of the most important options. The conversion required is related to established technology, since methanol is already produced in large quantities from natural gas, using a CuO/ZnO and other types of catalyst at 300 °C. In the case of gasifier or pyrolysis gas, the required reaction is:

$$CO + 2H_2 \rightarrow CH_3OH$$

and this proceeds exothermically under pressures of some 30–300 bar. Once again a shift reaction is required to give rise to the necessary $CO:H_2$ ratio; catalysts used include Al, Cr and Mn as well as Cu and Zn. In a wide range of US studies the costs of the finished methanol from biomass (municipal solid waste and wood) has been estimated at various levels from £2.74 to £8.70/GJ, quite an interesting range of possible costs for an automobile fuel and most attractive at the lower end. It seems quite sufficient to justify the increasing research interest in deriving transport fuel from this source.

(c) *Other liquid products*

Of all possible liquid products, ethanol and methanol appear to be the most readily derivable from biomass and hydrocarbon fuels with composition similar to that of petrol, the most difficult. However, the most encouraging aspect of the whole pattern of current research and development in the area is that so many of the potential opportunities have so far been little explored. For example, some catalyst mixtures have been found to yield higher alcohols along with methanol, and these come rather closer to paraffins in their energy content and combustion characteristics. A process has also been announced by Mobil for producing gasoline from methanol; though it appears to be accompanied by severe energy penalties, it nonetheless serves to illustrate that a wide range of options is likely to be open. Another approach is the much older Fischer–Tropsch process, which was extensively used in Germany during the Second World War to produce synthetic transport fuel and is now used in South Africa, in both cases with coal as the primary feedstock. The reaction is an interaction of hydrogen with carbon monoxide, like methanation, but, by a different selection of catalysts, longer-chain hydrocarbons are obtained:

$$nCO + 2nH_2 \rightarrow C_nH_{2n} + nH_2O$$

and

$$nCO + (2n + 1)H_2 \rightarrow C_nH_{2n+2} + nH_2O$$

The main disadvantage of the process appears to lie in the wide variety of the primary products (Brooks, 1942).

(d) *Other gaseous products*

As already mentioned, hydrogen can be produced from pyrolysis or gasifier gas by the reaction:

$$CO + H_2O \rightarrow CO_2 + H_2$$

and when CO_2 is scrubbed out, quite pure hydrogen may be obtained. This is, potentially, a source of hydrogen for premium uses, such as aviation fuel, for hydrogasification of more biomass, and for the production of ammonia and its salts for fertiliser use.

E Biological conversion processes

E1 ANAEROBIC DIGESTION

This process has been known for several centuries and the work of some of the early chemists in Europe makes reference to it during the late 18th and early 19th centuries. It consists of a microbial breakdown of biological material, typically sewage or livestock waste, evolving methane gas which is usable as a fuel, together with some carbon dioxide, while leaving non-digestible residues in the form of a slurry to be disposed of in other ways. The cities of Exeter and Birmingham in the UK operated digesters in the earliest years of the present century; in the case of Exeter it was for street lighting. Since then the process has become widely applied at sewage works and, for example, London has two very large ones each serving approximately 1.5 million people. By contrast, interest in anaerobic digestion of farm wastes has been spasmodic until recent years and restricted to a few individuals. The Second World War saw some increased interest, because supplies of alternative fuels were scarce, but throughout most of this century the availability of abundant and cheap fossil fuel has militated against the application of anaerobic digestion. On the other hand, in developing countries, especially in the East, fuel is often scarce and expensive and this has provided motivation for large numbers of small digesters to be run, especially in India, China, Korea, Taiwan and the Philippines. Most of these digesters are small, labour intensive, and involve no sophistification of design. Only now that several Western countries are again interested in the process of anaerobic digestion for livestock wastes, is any detailed attention being given to optimising the process concept and engineering. Indeed, it is now being widely realised that anaerobic digestion is applicable, not only to sewage and livestock wastes, but to any source or wet biomass such as grass, vegetable residues, algae and possibly also plantation crops grown specially as a feedstock.

The process is carried out by many different species of microorganisms acting together, no single bacterial species being able to carry out the entire conversion of a biomass component, such as cellulose, to methane.

The principle components of the feedstocks are polysaccharides, proteins, non-protein nitrogenous compounds, lipids, volatile fatty acids, salts and lignin. The lignin cannot be attacked by microbial action under anaerobic conditions and is left in the residues, together with most of the inorganic salts. The biological polymers such as polysaccharides, proteins and lipids must be broken down to simpler substances before they can be converted to methane and this calls for the action of many different types of bacteria. Some polysaccharides are especially difficult to degrade, as may be the case with cellulose and hemicellulose, especially if the fibres are lignified, the lignin serving to protect the polysaccharides from microbial and enzymic attack. Those feedstock components which can be broken down to simpler substances are eventually converted to simple organic acids, such as acetic, lactic, butyric and to ethyl alcohol and hydrogen. This is the end of one phase in the reaction sequence and the so-called 'methanogenic bacteria' take over from this point. They are of two kinds, one of which brings about a reaction between hydrogen and carbon dioxide to produce methane, the other decarboxylating acetic acid to yield methane. Examples of the first type are *Methanobacterium formicum* or *Methanobacterium ruminantium* and an example of the second type is *Methanobacterium methanca*.

Anaerobicity of the process is essential to its successful operation. The methanogenic bacteria themselves are the most prone to inhibition by oxygen and it has been stated that even a few molecules of oxygen per bacterial cell will prevent them from releasing methane. Among the other species of bacteria present there are usually many anaerobic species and many facultative aerobes, i.e bacteria capable of metabolising either in the presence or absence of oxygen, and these always tend to consume the oxygen in a livestock slurry and therefore tend to render it anaerobic as long as an excess of oxygen is not present. Therefore, when the feedstock, suspended in liquid, is confined in a closed tank, the conditions tend to become anaerobic and, so long as other conditions are correct, the bacterial species necessary for methane production will multiply. In practice it may take many days or weeks to get an anaerobic digestion process operating in the first instance unless an innoculum is added containing the correct mix of bacterial species from another digester, preferably one that has been operating with the same feedstock.

Other operating parameters include pH, temperature and constancy of temperature, stirring, and the presence or absence of inhibitory substances.

The pH is required to be in the range 6.5–8.0 and should be monitored, although, because of the good buffering capacity of the components of the feedstock, there is usually little need to adjust pH except during the starting up period.

The process can be operated in two temperature ranges, a lower or mesophilic range of from 10–42 °C and a higher or thermophilic range from about 50–70 °C. Between these two limits the reaction rate falls off and it appears that two quite different populations of bacteria are responsible for carrying out the process in these different temperature ranges. By far the commonest temperature range for practical digesters has been established at around 30–35 °C, well into the mesophilic range. The thermophilic population has a temperature optimum of around 65 °C, but little work has been done on the practical operation of digesters at this temperature, but Varel *et al.*, 1977 conducted experiments with cattle waste, bringing the detention time down from 20 days to 6 or less. The process economics may be adversely affected at thermophilic temperatures on account of the high energy demand for heating the feed unless efficient influent/effluent heat exchanges are installed, though the plant utilisation factor is much improved under thermophilic conditions.

Some form of agitation is usually required to optimise the rate of digestion by keeping the organisms well mixed with their subtrates. There are several types of unstirred digester but they usually suffer the disadvantages of low gas yield and/or long detention time in the equipment. Stirring can be conducted intermittently to save power. The plug flow digester (see below), in a design version studied by Hayes *et al.* (1979) which seems likely to offer acceptable reaction rate and yield, seems likely to dispense with agitation without detriment: at least one important factor appears to be the presence or absence of gas/liquid interface over most of the top surface of the reactor contents, since even well stirred digesters have encountered serious problems of crust formation at this interface when working on fibrous materials, resulting in outlet blockage and gas blow-out. In some designs, surface-directed jets of reactor contents combat this problem. Agitation may also be by gas circulation, rather than mechanical.

There are some problems of inhibition: sewage works digesters have been found susceptible to contamination by heavy metals (suggested critical concentration between 200–400 mg per kg of solids in the digester), to chlorinated hydrocarbons, (between about 10 mg and 2000 mg per kg of solids in the digester) and to synthetic detergents, particularly, aliphatic sulphonates and alkyl aryl sulphonates as they are both toxic and non-degradable. None of these problems need necessarily arise in digesting feedstocks other than sewage, which are not contaminated with industrial effluent or with domestic waste. Certain common inorganic ions have an inhibitory action when they reach critical concentration, for example sodium, potassium, calcium and magnesium can all have moderate inhibitory effects at from 1–5 gm per litre, which become severe at from 5–10 gms per litre, (see also McCarty et al., 1961). Sulphide can be inhibitory at even lower concentrations and, in view of the high nitrogenous content of livestock wastes, it is significant that ammonia is inhibitory at about 1.5 gm per litre. Cobalt is a desirable component, some 20 mg/l facilitating the process and this may well be linked with the suggestion that vitamin B_{12}, which contains cobalt, is involved in the mechanism of methane production by the methanogenic bacteria.

Although almost any form of fresh moist non-woody biomass may act as a feedstock to anaerobic digestion, the process design, the plant and the yield and composition of gas are in practice affected by the type of feedstock chosen since they differ in physical form, handling characteristics and in the percentage of the dry matter which is ash, or otherwise unavailable for digestion, and in the character and composition of digestible solids. The water content of the feedstock is also very important, since it determines reactor size and also has a crucial effect on thermal efficiency. All the water entering the digester has to be heated to digester temperature and hence, if a slurry is too dilute, there is insufficient thermal value in the gas produced to heat the large quantity of liquid to digester temperature and therefore maintain the reaction. In this case, outside heat has to be applied and the process efficiency is less than zero. Among livestock wastes the greatest difference is to be found between pig slurry and cattle manure, the two commonest forms of feedstock to digester installations at the present time. Pig slurry contains a high proportion of digestable matter, and before the gas needed to heat the digester is taken into account, the process may yield up to 65% of the available energy of the feedstock in the form of gas. On the other hand, cattle manure contains a high proportion of difficulty digestible fibrous material, some of which may be coated with lignin, which has already been subjected to some degradation by rumen bacteria. The most easily digestible parts of it have therefore been metabolised already leaving the most resistant component in the animals' waste. Therefore, the available materials for digestion in cattle manure are limited and the gas yield restricted to about 33% of the gross energy content of the feedstock. Animal manure contains a substantial proportion of suspended insolubles; these obviously affect the handling characteristics of the feedstock and therefore reactor design. Moreover, the presence of insoluble particles ensures that subtantial retention times are required, usually of the order of 10 to 60 days on account of the slowness of the hydrolytic phase of the reaction sequence. The bacteria attach themselves to the surface of the solid particles, particularly to digest the cellulose component, and this stage cannot be readily accelerated except by increased temperature, the rate bearing a relationship to the area of exposed surface of particles and therefore an inverse relationship to particle size. When plant matter is used as a feedstock the gas yield depends on the fibrousness of the particular feedstock, the amount of juice and the concentration of readily fermentable solubles which it contains. In the case of fresh grass some 60% of the gross thermal value of the feedstock may be recovered as gas. It should be noted, however, that plant feedstocks tend to act as a two component system, the readily diffusable low molecular weight substances being digested in a few days and the fibrous component requiring a far longer residence time.

The design of anaerobic digesters is extremely diverse, but most recent farm installations of properly engineered digesters have followed a similar pattern, known as the high rate slurry digester. Typically, the digester is a cylindrical tank receiving livestock slurry of 3–10% solids as feedstock. The digester is operated continuously, with additions of fresh slurry and removal of residues. The constancy of operating conditions is very important for maintaining the correct balance of microorganisms and any disturbance of temperature by more than 2 or 3 °C has a profound effect upon the bacterial population; recurrence of such disturbance may lead to gas evolution ceasing. With anaerobic digestion operated at higher temperatures, particularly in the thermophilic range, this susceptibility to temperature variation becomes more severe. The main thermal load in maintaining the temperature does not come about by heat losses through the walls of the digester, since these can be well insulated, but by the necessity to heat the large volume of incoming slurry and the consequent losses of heat in the outgoing residues. This heat loss may be reduced by introducing influent/effluent heat exchangers but the design of these is made difficult by the nature of the feedstock; when installed, they appear very expensive and still prone to blockage. Typically, the reactor is stirred intermittently either by mechanical stirrer or by a flux of the product gas. Most commonly the residues are drawn off as an unfractionated slurry and may be separated by a slurry separator into solid and liquid fractions. Alternatively, the reactor may be designed as to be able to draw off a fairly concentrated sludge with a high content of insolubles from the base and a free flowing liquid, containing only a smaller amount of very finely divided insolubles, from the upper part of the digester. This allows the draw-off of liquid and sludge to be separately controlled and the digester may then be operated 'with solids retention'; this may permit an increased content of active microorganisms to be retained within the system and therefore an increase of digestion rate. Limitations of this type of design are: (*i*) its thermal inefficiency, (*ii*) inability to handle vegetable material without comminution and (*iii*) proneness to crust formation.

There are, or have been, many variations upon this basic design. Some of these must be regarded as being superseded by the present-day high rate slurry digester. These include batch versions of the process, for example *unstirred slurry digesters* not fitted for continuous operation, and versions, known as contact digesters where the solids are separated from the residues after they emerge from the digester and after being partly de-watered, are returned to the reactor. This entails unnecessarily removing the microorganisms from their steady state environment and, therefore, upsetting their functional equilibrium.

There are also *solid form batch digesters* utilising livestock waste in France (Isman, 1979). These suffer from slow reaction rates, long detention times and the capital charges are likely to be high. On the other hand, there are no effluent problems the resulting residues being a quite desirable compost.

In the so-called *plug flow reactor*, there is a steady movement of the substrate in one direction through the reactor giving the advantage that substrates of relatively high solids content may be handled and the material which emerges as a residue has all been treated in the digester for approximately the same length of time, by contrast with the residues from completely mixed digesters, where part of the effluent always consists of fresh or nearly fresh feed. Until recently, plug flow reactors (Fry, 1975; and the design by the Gobar Gas Institute) exhibited slow reaction rate and high retention times. This appears to have been overcome in the work of Hayes *et al.* (1979) at Cornell University, in which detention times for cattle waste were similar to those for conventional digesters and gas yields were marginally improved. The plug flow digester, fully developed, could excel the conventional type by eliminating problems of materials handling and agitation, by allowing some reduction of capital cost and by improving thermal efficiency by allowing increased solids content. Therefore, an increasing number of workers are becoming interested in this design.

Packed bed reactors are a type of design that is of interest for effluents of low solids content. Ideally their feedstock contains organic compounds mainly in soluble form. The concept provides for a reactor containing solid packing materials, in the form of balls, hollow tubes etc., as is common in chemical engineering design for providing a surface area for contact. The principle involves the methane-forming organisms attaching themselves to the supports, thereby increasing the mass of active microorganisms in the reactor, by comparison with the conventional type. Very high reaction rates may be attained, since the feedstock is readily accessible to the microorganisms, and the active organisms are present in high concentration. Dead and living microorganisms are still released in the effluent, since they are rubbed away from the surface by the flow of material and, indeed, without this effect the reactor would become clogged. This type of reactor is clearly unsuitable for livestock wastes, in which there is a high content of insoluble matter, though it might conceivably be used for livestock waste after it has been subjected to a solids separation step. Finally, *multi-stage reactors* are based upon the knowledge that anaerobic digestion is a multi-stage microbial reaction, some of which, leading for example to the production of acetic acid, could well be carried out in a preliminary stage. Separation of stages involves increasing the capital investment in the reactor, but may be worthwhile, where the feedstock is insoluble; solubilisation can proceed slowly in the first stage, and solubles conveyed to a small, efficient second stage, which may even be based upon the packed bed principle. This concept is under investigation in Holland (Rijkens, 1979) and is illustrated in Fig. 2.

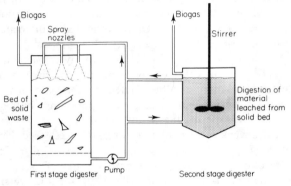

Fig. 2 Two-stage anaerobic digester

The use of mixed feedstocks has advantages provided their handling characteristics are not too greatly different. A single feedstock rarely has the optimum ratio of carbon to nitrogen for the mix of microorganisms. An optimum C/N ratio appears to be in the region of 10–30; livestock wastes tend to have a lower C/N ratio whereas fibrous materials such as straw, paper and municipal waste tend to have much higher ratio of C/N.

Livestock waste can be mixed with other organic materials containing less nitrogen, to give increased gas output. Indeed, higher utilisation of the available nitrogen by the microorganisms is likely, with more of the residue nitrogen appearing as microbial protein potentially separable for use as animal feed.

Biogas produced by anaerobic digestion comprises mainly methane and carbon dioxide. When pig manure is used as the feedstock the composition is typically about 65% methane, 35% carbon dioxide, giving a thermal content of 26 MJ/m^3, and it is usually a little higher when poultry manure is used (about 70% methane, 28 MJ/m^3) or lower when cattle manure is used, being about

57.5% methane (23 MJ/m^3). When the gas is used on farms, at sewage works or other on-site situations, for little or no attempt is usually made to remove the carbon dioxide. However, the carbon dioxide may be removed if desired, when it is worth the cost of doing so, by means of alkali scrubbing, scrubbing with pressurised water or by absorption onto zeolites or into an organic solvent such as ethanolamine. As projected digesters become larger, with the gas being distributed to substantial areas or being used for industrial purposes, the relative importance of carbon dioxide removal to increase the thermal content, increases. Hydrogen sulphide is a minor component of biogas, variable dependent upon the feedstock used, but typically about 0.1% of the volume. Its combustion gives rise to sulphur dioxide, which may cause corrosion of engines and pipes, and hydrogen sulphide itself is toxic in quite low concentrations. Removal of hydrogen sulphide is possible by scrubbing with water under pressure and then exposing the gas to iron filings, or by oxidation with anthraquinone disulphonic acid (Holmes–Stretford Process) to give elementary sulphur. Hydrogen and nitrogen or ammonia, have sometimes been identified as minor, sporadic components of biogas.

Yields of biogas may be conveniently expressed in terms of m^3/t or m^3/kg of feedstock. This approach may be sufficient when considering a digester to work on one constant feedstock. However, feedstocks vary in moisture content and ash content and therefore, to put yield figures on a common basis, it is common to express yields as m^3/kg or m^3/t of dry solids or of dry organic solids, i.e. the moisture and the ash being disregarded. Theoretical yields calculated from basic chemical principles would be 800 l/kg, from carbohydrates, 700 l/kg from proteins, 1200 l/kg from fats. By contrast, experimental yields of gas have been found generally to be about 200–350 l/kg for cattle manure, about 400–500 l/kg for pig manure and 600 l/kg has been reported using sewage sludge. By comparing the highest and lowest of these figures the maximum and minimum gross energetic efficiency may be calculated. A low figure, falling within the range for cow manure, is 0.3 m^3/kg of gas with thermal capacity 22 MJ/kg, which corresponds to 6.6 MJ/kg organic solids and a 33% gross yield of the energy in the feedstock in the form of product gas. On the other hand, an optimum gas yield would be represented by 0.6 m^3/kg organic solids of gas with 28 MJ/kg, and this corresponds to gas with a thermal value of 16.8 MJ/kg organic solids, or 84% gross yield of the feedstock energy in product gas. Typical values for mixed substrates lie between these two extremes. An important economic parameter is the yield of heating value in usable fuel per unit of incoming feedstock. A proportion of the product gas or other fuel, has to be used to heat the feedstock entering the digester and the absolute amount of heat is almost independent of the concentration, composition and digestibility of the feedstock. Calculation shows that a feedstock containing 8% w/v or oganic solid having a gross energy content of 20 MJ/kg contains 1.6 MJ/l; at a 60% gross thermal deficiency this yields 0.96 MJ/l as gas. The theoretical heating requirement, if the slurry has to be heated through 25 °C, is 0.105 MJ/l but in practice, this heating requirement must be doubled to allow a degree of boiler inefficiency and for heat losses in the pipework and through the digester walls. Under these conditions, therefore, some 25% of the product gas is required for heating the digester and feedstock. If the concentration of the incoming slurry should fall to only 4% of organic solids, and its gross energy content, therefore, to 0.8 MJ/l the heating requirement per unit of incoming slurry remains the same while the heat available from the feedstocks is halved, and therefore something approaching 50% of the product gas must then be used to heat the digester and feedstock to reaction temperature. Moreover, should the efficiency with which the organic solids in the feedstock is converted to biogas fall to 33%, as it may do in the case of cattle manure, then the percentage of product gas required for digester heating rises to 45% at an organic solids content in the feed of 8% w/v, or 90% at a feedstock concentration of only 4% organic solids in the feed. These considerations have a very marked effect upon the cost of energy in the product gas and for example, would make digestion of cattle slurry at the lower concentration of only 4% w/v entirely uneconomic. Another factor governing economics of digester operation is the residence time, which largely governs the gross output of a digester of a given volume. Here again cattle manure suffers a disadvantage, since it is widely considered to require a residence time of about 20 days.

Finally, careful consideration must be given to planning the utilisation of biogas. Experience has shown that where it is produced on a farm it is most often convenient to use it for electricity generation, all the biogas generated being burned in the engine, with waste heat being recirculated to heat the digester; any surplus waste heat would be available for heating the farmhouse or, where necessary, animal houses. The principle problems arise because electricity utilisation on most farms peaks at certain periods of the day and troughs at night, while gas production is at a continuous steady rate. Also, uses are not usually to hand for all the waste heat. Another possible scheme involves farmers pooling their resources to a single digester and sharing the use of the energy generated.

Other possible uses for biogas are for domestic heating and industrial energy in rural areas; the heating demand is seasonal and although the seasonality roughly coincides with the availability of feedstocks, the winter is also the time of year when digesters are least efficient thermally and it is doubtful whether, at the present time, the capital intensive cost structure of anaerobic digestion projects would allow seasonal use with long idle periods in the summer. Perhaps one of the most attractive possibilities is the neighbourhood digester serving an area for several miles round, where the energy can be utilised in a planned and integrated manner for steam raising, space heating and electricity generation in rural industries, or even industries on the edge of towns and cities. In such a case economics could be improved by the use of straw or wood for digester heating, and this would also eliminate the inconvenient seasonal variation of net output. It has also been suggested to employ biogas for propelling farm vehicles.

E2 PRODUCTION OF ETHANOL (ETHYL ALCOHOL)

The production of ethanol by yeast fermentation of simple sugars has been known throughout history as the means for production of alcoholic liquors. Fuel alcohol is a recent idea but ethanol has been produced as a chemical, as surgical spirit and for solvent purposes for several decades.

From the standpoint of energy production, the ethanol route suffers from the disadvantage that free sugars are required as substrate, and feedstocks with a high content of free sugars, or of polysaccharides readily hydrolysable to free sugars, are normally in demand for higher value purposes than fuel production, such as food, feed and for high value fermentations, such as the antibiotic industry (see chapter, 5 section B). Successful production and utilisation of ethanol as fuel therefore depends upon a means for producing sugary substrates in large volumes by cheaper than normal means and upon the relatively high demand and high price for convenient liquid fuels comparable to petrol suitable for fuelling automotive engines.

The basic biochemistry of ethanol fermentation consists of a reaction sequence leading to the overall conversion:

$$C_6H_{12}O_6 \rightarrow 2C_2H_5OH + 2CO_2$$

this reaction proceeds through a dozen or so enzymically catalysed stages, with a characteristic sequence of intermediates known as the Embden-Meyerhof Pathway (Baldwin, 1952). The overall sequence releases energy in a metabolically utilisable form which sustains the life and growth of the yeast (adenosine triphosphate, ATP). In order to carry out the reaction the yeast requires mineral and vitamin nutrients, but these are normally present in the biological substrates used. The industrialisation of the process, starting from whole plant material containing sugary sap, calls for equipment to express the sap (e.g. sugar cane rolls), seed fermenters to contain modest quantities of the feedstock while the yeast is multiplying its numbers and larger fermenters in which the major amounts of feedstock are transformed while the yeast which is 'seeded' into it continues to multiply. The sugary substrate is normally at about 12% w/v and is fermented for 36 hours. With substrates deficient in nitrogen, ammonium salts have to be added. The organism employed is *Saccharomyces cerevisiae*. Finally, equipment is required for purifying the alcohol from the rest of the fermentation mixture by separating the solids and by distillation, and, if very pure alcohol is required then special measures must be taken to remove the final few per cent of moisture that cannot be separated by distillation alone.

Where the feedstock is rich not in free sugars, but in polysaccharides, a hydrolysis step must be performed, either under mild acid conditions, or with an amylase (starch-splitting) enzyme. Starch is the most likely easily hydrolysable polysaccharide to be encountered, in crops such as cassava, grain or potatoes, but inulin, which yields fructose on mild hydrolysis, also occurs in potential energy crops such as the Jerusalem artichoke and chicory.

Brazil has taken a unique approach towards independence from imported oil by legislating their own National Alcohol Programme in November 1975 and backing it with massive resources. The programme set an initial objective of 3M m^3 of anhydrous alcohol per year in order to achieve a 20% addition to all gasoline during the 1980s, the feedstocks being sugar cane and cassava. Sugar cane was used first because alcohol production from molasses had been practiced for some time for non-fuel purposes, and the technology is easier, not calling for hydrolysis. However, sugar cane production was recognised to be limited by the availability of irrigated land of good quality and cassava was needed for its ability to grow on much poorer soil. Three types of distillery were recognised among the 120 new ones projected; (*i*) *the annexed distillery*, which is attached to an existing sugar factory, sharing services and operating on cane juice or molasses, (*ii*) *the central distillery*, which is a separate distillary serving several factories within an area and operating on cane juice or molasses and (*iii*) *the autonomous distillery*, which has its own estates for supplies of crop and operating directly on sugar cane or cassava, (Jackson, 1976). Scale of operation is from 20 000 up to a maximum of 400 000 m^3 of anhydrous ethanol per year. About 70 l of ethanol/t is derivable from sugar cane, 150–170 l/t from cassava, although cane has the advantage of higher yield/ha and of providing process fuel in the form of bagasse.

For production from sugar cane the steps are:
(*a*) Efficient extraction of fermentable solids by diffusion.
(*b*) Clarification.
(*c*) Pasteurisation or sterilisation.
(*d*) Fermentation, using the Melle-Boinot method in which yeast is recycled to keep up the active cell concentration.
(*e*) Centrifugation, settling or filtering.
(*f*) Distillation, often in two stages of concentration, to 95% ethanol.
(*g*) Distillation of azeotrope, to yield anhydrous ethanol feedstock.

When the feedstock is cassava, the stages are:
(*h*) Root trimming, giving roughly 2% loss, peeling and washing.
(*i*) Grinding, homogenising and heating to break starch grains.
(*j*) Saccharification with amylase, commonly derived from a microbial source, such as *Aspergillus niger* at 50 °C for 60 hours, in a separate, aerobic, fermentation process.
(*k*) Filtering.
(*l*) Fermentation, as (*d*) above.
(*m*) Steps (*e*), (*f*) and (*g*) as above.

The yield of ethanol from cassava represents approximately 40% based upon cassava dry matter, or one l of ethanol from 6 kg fresh cassava at 66.7% moisture. With a maximum cassava yield of 50 t/ha after say, 18 months growth, the harvest from one hectacre yields a maximum of 8300 l of ethanol, or about 5500 l/ha/yr.

It seems quite unlikely that similar processes will be operated on a large scale in Europe, or in any country with substantial pressures upon its agricultural land resources, for reasons given in chapter 5, section B, in brief, the tropical crops are not applicable and the nearest temperate equivalents cannot approach their performance and costs. One possible source of sugar-rich solids, would be plant juices from crops grown on a large scale in moist temperate climates as combined fodder and energy crops and subjected to fractionation by juice expression (chapter 5, section E), although this source would suffer the disadvantage of low concentration.

In the European agricultural and energy scene, it is more likely that sugars for ethanol fermentation could become available from hydrolysis of cellulose than from sugary or starchy crops. Agricultural residues are predominantly cellulosic, and the same is true of the more promising energy crops for temperate, developed countries, a feature which is all the more emphatically true because in the first instance, at least, such crops could only be allocated to the poorest ground. Either non-woody energy plantation crops or, indeed, forestry products, are potentially available for ethanol production after hydrolysis of their cellulosic component to sugar, and the same is true of the cellulose in solid municipal waste. At present the most practicable route is likely to be high-temperature, short-time hydrolysis with dilute mineral acid, as developed by Converse *et al.* (1971). In this process, the cellulosic material is suspended in dilute sulphuric acid (about 1%) and passed through a continuous tubular pressure reactor at 200–300 °C. The shorter the residence time and the higher the temperature, the better the yield of sugar, since the hydrolytic step is accelerated by temperature to a greater extent than side-reactions which consume glucose. Under any given set of reaction conditions the glucose yield rises to a peak after a characteristic time for those conditions, and then falls away as sugar is destroyed without being replaced, giving curves such as those in Fig. 3. Yields amounting to over 60% of the cellulose fed can be achieved at temperatures over 200 °C, but the residence time falls to a few seconds only, becoming difficult to monitor and control at the highest temperatures and shortest times. Fuel consumption is high, probably of the order of 10 GJ/t of sugar, making it very likely that the total conversion of cellulose to ethanol would consume more energy than the gross thermal value of its output. Lignin and other undigested material remains as a solid residue with a thermal content higher than average biomass and potentially convertible into a solid fuel. The process can be more complex than envisaged above if the feedstock contains a substantial proportion of the more readily hydrolysable hemi-cellulose, since these may call for a preliminary stage of hydrolysis at a lower temperature if overall sugar yield is to be maximised.

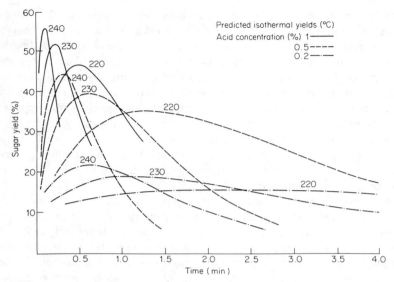

Fig. 3 Sugar yields in acid hydrolysis of paper. (After Converse et al., 1971)

Attempts have been made, and are continuing, to provide an alternative to this costly and energy-intensive approach, by developing enzymic methods for cellulose hydrolysis. After many years of effort, this is still only at the research stage, cellulose being much more difficult to degrade than starch on account of the different configuration of its glucose/glucose linkages (1:4 β- rather than the 1:4 α-links in starch), the nature of the fibrous bundles in cellulose providing major steric hinderance factors against the approach of the enzyme to its site of action, the cellulose fibres often being coated with lignin, and the cellulose enzyme systems being more complex and interdependent in their mode of action. The present position, therefore, is that no practical enzymic processing system exists for converting ligno-cellulosic tissue — the most abundant form of biomass — into sugar, at a cost suitable for fermentation to ethanol as fuel. There is little doubt that European countries and the EEC will keep both chemical and enzymic hydrolysis methods under review in case any breakthrough in costs and methods should emerge.

E3 OTHER POTENTIALLY AVAILABLE BIO-CONVERSION PROCESSES RELEVANT TO FUEL APPLICATIONS

Anaerobic digestion and fermentation to ethanol are the only two bio-processes that are being seriously worked upon for fuel applications, but this is not necessarily the limit of involvement of biotechnology in the field. The metabolism of plants and bacteria is extremely diverse and impinges upon most classes of chemical substance. Now that so much work is being directed towards manipulating the gene structure of organisms the techniques involved could well be applied to fuel-producing reactions; these, together with the well-established methods of strain-selection and the slowly emerging techniques for culturing metabolically active plant cells in continuous culture, embody the possibility that any biological system that is now capable of producing a fuel chemical in minor yield, could in future be induced to give the same chemical as a principal metabolic product. Naturally, some of these conversions are less credible than others as future commercial processes.

An alternative gas-yielding reaction to bacterial methane production would be the bacterial reaction:

$$C_6H_{12}O_6 + 6H_2O \rightarrow 6CO_2 + 12H_2$$

Although this appears superficially very attractive, enabling more than 99% of the thermal value of glucose to be conserved in H_2, Thauer (1977) has shown that the overall change is associated with insufficient free energy change to drive the reaction and to allow the fermentation organisms to derive metabolic energy from it; these considerations limit production of hydrogen from carbohydrates to about 4 mol/mol of hexose and hence to around 33% of the theoretical yield. Alternatively, hydrogen may eventually be produced via the photosynthetic systems of higher plants or algae by:

$$2H_2O \xrightarrow{\text{light}} O_2 + 2H_2$$

in a biotechnical energy conversion process not involving a biomass stage (biophotolysis), as discussed by San Pietro (1977). The blue-green algae often used in such experimental systems, such as *Anabaena cylindrica*, are well able to utilise the hydrogen they generate for nitrogen reduction, to produce ammonia which although not a convenient fuel, is a major commercial product normally requiring large fuel inputs for its manufacture.

Production of methanol from methane is also conceivable by blocking the onward metabolic reactions from methanol in methane-utilising bacteria (Foo and Heden, 1977). Other possible routes are the microbial production of fats and fatty oils (Ratledge, 1976) or of hydrocarbons (Tornabene, 1977), although to date the levels of production in microorganisms of acyclic non-isoprenoid hydrocarbons remains unimpressive.

[3]
Crop wastes

A Range of sources and availability of crop wastes

Most agricultural crops are grown for a particular portion of the plant, which has a distinct market value. Hence, in conventional terms the yield figures quoted for wheat or barley for example, apply only to the grain portion, being typically of the order of 4–5 t/ha in temperate climates. In relation to this, straw yields most commonly vary from 100% to 60% of the grain yield, and hence from 5 to 2.5 t/ha, while the unharvestable stubble and underground portions of the plant have been estimated to amount to about 2.5 t/ha as stubble and 2 t/ha as roots, all expressed as dry matter (Russell, 1977). On this basis, typical temperate cereal crops appear to be yielding from 11 to 14.5 t/ha/yr dry matter as the total product of their photosynthesis, of which only about 35% is readily harvestable and saleable off the farm. It is not even clear whether Russell's figures include the chaff and fine leaf fraction of straw which is largely dropped back onto the ground during combine harvesting. Together with cavings, shrivelled corn and weed seeds, this fraction has been estimated at 10–15% of total crop weight and is largely left on the ground after combine harvesting and baling, together with 4–5% of the grain, which the combine has failed to collect (Lucas, 1978; Vahlberg, 1978). Biomass energy schemes, as opposed to conventional agriculture, are concerned potentially with the total output of organic matter from a crop, provided it is harvestable. In the particular cases of wheat and barley, no method exists for harvesting the stubble and roots, so that this portion is unavailable for energy purposes; it is also quite likely, as indicated by Russell, that cereal growing land is rather dependent upon this annual input of organic matter to maintain a minimal humus content. Certain other plants do have potentially harvestable underground parts, even though they are not normal agricultural root crops (e.g. *Pteridium*, see chapter 5). However, this example serves to illustrate the attitude to crop wastes that is required in order to develop the biomass/energy concept.

A1 RESIDUES DRY ENOUGH FOR COMBUSTION

Straw, from cereals, and similar stem-wastes from other crops, constitute the largest source of plant waste matter arising in agriculture. They are distinguished from other crop wastes by relatively low content of moisture (e.g. about 14% moisture is commonly found with cereal straw). Hence, these wastes, as fuels, are ready to be combusted or thermally processed without further dehydration and processing, in contrast to wastes from vegetable-growing, which typically contain from 78 to 84% moisture in which case, unless they can be conveniently sun or air-dried, they lend themselves most readily to gas production via anaerobic digestion.

Apart from wheat and barley, these relatively dry stem-wastes include rye and oat straw, rice straw and maize stover, and dry residues from rape, colza, dry peas and beans grown for fodder. Consideration of tropical crops leads to consideration additionally of sugar cane bagasse (the fibrous material left after sugar extraction), the woody stems of cassava, wastes arising in the production of commercial fibres, such as jute, hemp and sisal and bush prunings on tea plantations. Sugar cane bagasse is not very dry as it arises, but is usually pressed to about 50% solids content.

Other dry residues arise, which are not stems, but which may be utilised in the same ways, such as nut shells, for example almond shells, which are sometimes used to fire boilers in Mediterranean countries, coconut shells and the woody wastes which arise within agriculture when old vineyards or orchards are grubbed up for replanting.

Naturally, a proportion of these fairly dry woods are utilised at present, wherever it is particularly convenient to do so, or where fossil fuel is especially scarce or expensive, wastes may have already been pressed into use. This has happened quite widely, for example, with sugar cane bagasse, which is often used to fire boilers to provide energy for the sugar cane processing plant itself. In the majority of cases, however, these wastes have not been used for reasons connected with the much greater convenience of liquid fuels, the larger, specially designed furnaces required for burning crop wastes, the difficulty of arranging feeding of crop wastes, the labour involved in collecting the wastes and storing them, or, in some cases, the lack of a nearby market for the fuels.

Many of these wastes would be freely available as fuels, so that utilisation merely awaits the development of economic systems for handling and burning them, so as to provide useful energy, in places where it is needed, at prices to compete with prevailing fossil fuel prices, especially oil. It is a situation which appears to call for action by the governments to encourage and inform people concerning economic opportunities to use these materials and, in some cases, to provide an organisation and infrastructure necessary to make such utilisation viable. On the other hand, a substantial proportion of cereal straw is generally in demand for livestock bedding and other agricultural applications and in many areas, including several departments of France and much of the cereal land of western Canada, (Coxworth, 1978), the ploughing-in of straw is considered necessary to maintain humus levels, enabling the soil to retain moisture and establishing a crumb-structure. In these areas, the harvesting of the straw would be considered disadvantageous and it is rather unlikely to be applied to energy purposes. In the UK, it has been shown that grain yields are increased by ploughing-in the straw in areas where the soil humus has become, or is naturally, depleted (Johnston, 1978).

On a worldwide basis, the production of dry agricultural residues represents a great deal of energy. World combined production of wheat, maize, barley, oats, rye and rice straws, together with sugar cane bagasse, exclusive of any figures for China, probably amount to some 900 Mt/yr of dry matter, or approximately 360 Mt of oil equivalent. The corresponding figure for the countries of the European Communities is 45 Mt of oil equivalent, though in this case no sugar cane bagasse is included.

A2 WET AGRICULTURAL RESIDUES

Wet agricultural residues generally comprise green leafy material or moist stems of vegetable crops, the type being very dependent on geographical area. Vegetable crops are more diverse and preferences of food choice tend to be much more local than in the case of staples. This makes it impossible to consider the global situation at a glance, and most of what follows relates to the vegetable-cropping pattern of NW Europe.

Two large commodity items make an important contribution to potential wet waste arisings, sugar beet and potatoes. Sugar beet, wherever there is a significant industry, tends to be the most important, both in terms of the quantity of residue arisings and the prospects for harvesting them for energy purposes.

The waste arisings from sugar beet are in the form of the green tops, together with part of the crown of the beet, which is cut off at the same time. The waste arisings from the potato are in the form of the green top of the plant, or haulm. The weight of it is in much lower ratio to the weight of the crop than in the case of sugar beet tops and, so long as present practices prevail, they will not be readily available for energy purposes. The haulm tends to be senescent when the tubers are harvested and their destruction is often completed by use of a chemical spray, such as sulphuric acid. However, there appears to be no real obstacle to the harvesting of potato tops if they were required, so they may be included as a potential energy feedstock.

Other wet wastes that could contribute to energy comprise the Brassicas, especially Brussel sprouts, cabbage and savoys, cauliflower and broccoli, and the peas and beans. Waste arising from Brassicas include stalks, outer leaves and trimmings, but there is also a contribution from malformed crop, since deformed cauliflowers, or cabbages which fail to form hearts, cannot be marketed. The waste from peas and beans comprise the haulms and, in some cases the shucks. Other waste vegetable matter arises from root crops, such as carrots, or from horticulture, especially tomato haulm discarded at the end of the season. However, although these wastes may be significant to some individual large growers, they are quite insignificant quantitatively in national terms.

Finally, agricultural enterprises are sometimes involved in the further processing of their own produce. Vegetable-growers sometimes pack for sale through supermarkets, necessitating trimming operations, since, for example, some supermarkets take only the curd of the cauliflower, without leaves, and this generates more waste at the farm. Vegetable-growers sometimes also freeze their produce, necessitating blanching, which produces a dilute liquid effluent containing organic material. There is, indeed, a generally narrow dividing line between waste arising in agricultural operations and waste arising in processing. In warmer countries, olive processing waste or palm waste, from palm oil extraction, are significant liquid effluents. Similarly, although in temperate agriculture, fodder crops rarely give rise to any discrete and collectable solid wastes, liquid effluent is lost in quite large quantities from silos in which fodder crops are ensiled for winter use, and this effluent is very amenable to anaerobic digestion. All these are potential sources of energy, but detailed treatment is only given below to solid, potentially collectable residues, with potential to make some significant contribution, quantitatively, to energy supply.

B Quantities and collectability of crop wastes

B1 CEREAL STRAW

The statistic which is generally available for any given area, relative to cereal production, is grain yield; straw yields are only occasionally measured and reported, as they are clearly of much less general interest. Therefore, estimates of straw production often depend upon knowledge of grain/straw ratios, which may be applied to grain statistics to derive estimates of straw production. Typical values for cereals are wheat, 1.23; barley, 1.45; oats, 1.16 and rye, 0.70. For maize, a cob/stover ratio of 1.00 is quoted by Meriaux *et al*., 1977; for rice, a Stanford Research Institute, 1976 report gives figures which indicate a ratio of grain to harvestable straw of 0.78, whereas the ratio for rape and colza is generally in the region of 0.83. These ratios can be used to calculate the total yield of harvestable dry stem residues in any area for which grain production statistics are available and they provide a useful guideline. However, in practice straw yields are undoubtedly influenced by factors other than the pattern of cereal crop species. In particular, variety selection can be significant. For example, among winter wheats Maris Huntsman gives a grain/straw ratio of 1.3 compared with Maris Fundia, at between 2.2 and 2.7. Hence, any area in which very short stem varieties were prevalent would show a major departure from typical straw yield patterns. A French study (Meriaux *et al*., 1977) has shown that lower rates of insolation diminish the yield of grain more than the yield of straw; consequently higher latitudes can be expected to exhibit lower grain/straw ratios than lower latitudes. Incidence of cold weather immediately prior to flowering has a similar effect; occurrence of water deficit has a variable effect, dependent upon its

CROP WASTES 53

timing. Not surprisingly, the different impacts of genetic, climatic and cultivation factors are difficult to relate to actual national straw yield figures. For example, in the UK a straw survey (Hughes, 1977), which was supported by direct observation of straw-baling operations, gave results leading to an overall grain/straw ratio in UK cereal farming close to 1.40. Climatic factors seem unlikely to be responsible for this high figure, which may be more related to the particular varieties selected for sowing.

Yields of both straw and grain are subject to variations, through climatic factors, from season to season, and this is undoubtedly responsible for much of the variations that can be seen between straw yield estimates published at different times. For the UK the most probable range of values for national cereal straw yield is 11–13 Mt/yr, for Germany 19–23 Mt/yr, for France 26–30 Mt/yr, for Italy 7–9 Mt/yr. In European countries most of these figures are accounted for by wheat and barley, but with rye and oats providing a more significant contribution in Germany and France than elsewhere. In the case of Germany, approximately equal quantities of wheat straw and barley straw are produced; in France wheat straw predominates, being over 60% of the total, whereas the UK produces far more barley straw than wheat straw. These differences are important in that they affect the range of competitive uses for the straw and hence, when it is offered for sale, its market price.

If all cereal straw could be collected and used as fuel its gross thermal value would be approximately 360 000 TJ/yr in Germany, 480 000 TJ/yr in France, 140 000 TJ/yr in Italy and in the UK 200 000 TJ/yr. In practice these values are diminished by about 12% to take account of the heat needed to drive off the residual moisture when the straw is burned. Bone dry straw has a thermal content of approximately 17 MJ/kg, compared with typical air-dry straw of some 14% moisture, at 15 MJ/kg.

B2 SUGAR BEET TOPS

Sugar beet is a major European crop, being planted to some 420 000 ha in Germany, 530 000 ha in France, 240 000 ha in Italy and 210 000 ha in the UK. Yields vary from 4.7 t of sugar/ha in the UK to around 6.7 t/ha in several other European countries, corresponding to typical beet yields of 34 t/ha and 48 t/ha respectively, with a mean at approximately 41 t/ha. The yield of tops is very similar to the yield of beet, being put at 0.95 of the beet yield by several experts (Ader and Buck, 1979). The equipment for topping usually also removes a slice of the crown of the beet, so that the average yield of tops may be taken as approximately equal to the beet yield. Typical solids content for the tops is 16% so that a 5.5 t/ha yield of dry matter in tops is applicable as an average for the UK, 7.6 t/ha elsewhere in Europe. These figures lead to substantial overall yields; 1.15 Mt/yr in the UK, 3.00 Mt/yr in Germany, 4.03 Mt/yr in France and 1.55 Mt/yr in Italy; at an estimated 17.5 GJ/t of dry matter the gross energy in these residues is equivalent to approximately 20 000 TJ/yr in the UK, 50 000 TJ/yr in Germany, 27 000 TJ/yr in Italy and 70 000 TJ/yr in France.

In practice it is recognised that some of the tops are not recoverable, on account of damage done by the harvesting machinery. Single-row harvesters may damage half of the tops by crushing them into the soil, while even double or triple row harvesters may still damage about 25%. Six-row machines, fitted with sensors to reduce the amount of the beet crown sliced off, may eventually be able to achieve a 90% recovery of tops. However, Nuttall (1979) reports on experiments in which 40 t/ha of fresh tops were harvested, and this yield had clearly been computed after harvesting losses. This illustrates that very good yields, are obtainable at the present time. The equipment used was:

(*i*) A six-row Moreau topper with chopper blower attachment, which blows directly into trailers to minimise soil contamination.

(*ii*) A three-row Standen Multibeet root lifter.

(*iii*) A 7-8 tonne high lift tipping trailer carting tops.

It should be noted that the yield of tops varies with harvesting date: Nuttall (1978, 1979) found that 41.7 t/ha could be harvested on 13 September, but that this diminished continuously as the season advanced, until only 24.8 t/ha could be harvested at mid-December, the average over the season being 31.7 t/ha. These losses that take place as the crop stands in the field may be inevitable, since beet harvesting must take place throughout the 'campaign' to keep the factories fed with beet; the only option to avoid them would be to cut the tops well in advance of lifting the beet.

B3 POTATO HAULM AND REJECT TUBERS

The areas planted with potatoes annually are approximately 400 000 ha in Germany, 300 000 ha in France, 180 000 ha in Italy and 230 000 ha in the UK. Yields are in the range 26–28 t/ha in most countries, but only 16–17 t/ha in Italy. At a haulm/crop ratio of 0.176 (Monteith, 1977), this corresponds to a mean of 4.75 t/ha of haulm which, at an estimated mean dry matter content of 23%, amounts to 1.09 t/ha of dry matter. These figures do not disclose the distinction between early and main-crop potatoes. These may be expected to differ; the likelihood is that early potatoes would be found to have a higher ratio of haulm/crop, since the additional growth during the latter part of the season will be mainly in tubers. The overall average of 1.09 t/ha of dry matter converts to a 7600 TJ/yr of gross energy for Germany, 5700 TJ/yr for France and 4400 TJ/yr for the UK; Italy, which shows a lower average yield, may or may not show altered haulm yield accordingly, assuming that it does, the gross energy in the haulm will be 2100 TJ/yr. Holland, with 170 000 ha planted, may be presumed to produce haulms with a gross energy content of 3200 TJ/yr.

The practicability of gathering haulms is clearly shown by the 34% of the UK crop which is at present 'topped' and the haulm removed mechanically, although half of this is treated with defoliating agents as well. Some 55% of the crop is treated with defoliating agents, but not topped, whereas 11% are neither defoliated or topped. There is clearly every possibility of gathering these haulms if there was a need to do so and the present practices, which largely disregard the haulm, have doubtless evolved because, containing high alkaloid levels, it is useless for animal feed. It may therefore be concluded that quantities of haulm approaching the levels calculated above could be made available for energy purposes as a result of altered harvesting practices, though they are not available for collection under present circumstances.

Reject potato tubers are stated by Francis (1976) to amount to 3.75 t/ha in the UK, but as these appear to be generally quite efficiently utilised as annual feed, they will not be considered further.

B4 HAULMS AND SHUCKS FROM PEAS AND BEANS

The total quantities of waste arising in England and Wales from pea and bean crops are estimated in Table I. Approximately half are attributable to green peas grown for processing. The total of about 9000 TJ/yr is small in relation to other sources of agricultural waste; the prospects for actually collecting them to be used for energy purposes differ in a complex fashion among the different classes of crop. Significant features making it inconvenient to collect the waste, include the sheer speed with which peas for processing must be harvested and moved to the factory, the design of harvesting machines which are built to deposit mutilated wastes on the ground, the need to avoid compacting the soil with unnecessarily heavy machinery and the pressure on workers and equipment at harvest time.

Green peas for processing have been harvested for many years by first cutting the vines and then windrowing them, ready for mechanical vining using a mobile viner, which strips the pods and collects them in a trailer, whilst the pods and haulms are returned to the ground. It is possible to collect this waste by means of an elevator and secondary trailer; indeed, this used to be done and the residues ensiled for animal feed, but it is no longer considered economic. Moreover, mobile viners are currently being replaced rapidly by mobile pod-pickers, which strip the vines, leaving the stems in the ground, separate the peas to a trailer, returning much-mutilated pods and haulm fragments to the ground. Collection of waste from pod-pickers would be possible after machine modifications, but as these machines do not harvest the entire haulm in the first place, this option is not very attractive.

By contrast, peas for harvesting dry give rise to a readily collectable waste, and there would be no difficulty in collecting haulms (but not shucks) from peas grown for market, since they are hand picked. Beans for stockfeeding also give rise to waste that is readily collectable, but beans grown for human consumption are generally harvested mechanically and securing the waste is subject to the same difficulties which apply to peas for processing. The overall effect of these different situations applying to different categories of peas and beans is that only some 33% of the total waste generated can be considered collectable under present circumstances, though a further 23% could become collectable, subject to modifications to harvesting machinery.

B5 BRASSICA WASTES

The total quantities of wastes arising in England and Wales from Brassicas are shown in Table I. Well over half are attributable to Brussels sprouts, but wherever these are mechanically harvested, leaving only the stalks in the ground, a separate harvesting operation would need to be introduced for the stalks, which are otherwise ploughed in. The same is true of hand-stripped Brussel sprouts. Some mechanical harvesters (Jamafa) incorporate a system for handling whole plants, including leaves; all residues being deposited on the ground; collection of these would require the use of an elevator and secondary trailer. Some of the crop (about 15% in England and Wales) is cut and carted in bulk trailers to a central machine-stripping shed; as with other harvesting methods, removal of the leaves by hand precedes the harvest, so that only stalks would again be available. The total weight of stalks that could be available in England and Wales, after introducing either a second harvesting operation or modifications to harvesting machines, amounts to 70 000 t/yr dry weight, or 1200 TJ/yr. However, the stalks are tough and fibrous; they could not be fed to an anaerobic digester without being disintegrated first, and in view of the modest quantities involved, this may not be worthwhile.

Table I Pea, bean and Brassica wastes in England and Wales

Vegetable	Cropped area (ha), 1977	Estimated residue arisings (t dry matter/ha)	Residue yield (t dry matter, to nearest 100 t)	Estimated gross energy of residues (TJ/yr)
Peas and beans				
Green peas, for processing	55 644	4.93	274 100	4660
Peas for harvesting dry	36 994	2.58	95 400	1620
Green peas for market	4310	3.35	14 400	250
French and runner beans	13 400	4.28	57 400	980
Broad beans	5610	4.28	24 000	420
Beans for stockfeeding	38 090	2.06	78 500	1370
Total peas and beans	154 048	3.53	543 800	9520
Brassicas				
Brussel sprouts	14 798	5.28	78 100	1330
Cabbage and Savoys	18 890	1.32	25 300	430
Cauliflower and broccoli	12 499	2.61	32 600	560
Total Brassicas	46 187	2.95	136 100	2310

Wastes from cabbages, Savoys, cauliflowers and broccoli that are harvested for market are generally unavailable for energy because the crop does not mature uniformly and is unsuited to mechanical harvest. Serial cutting and trimming is performed by hand, so most of the outer leaves have decomposed by the end of harvesting; the stalks are too short to be considered for harvesting separately.

CROP WASTES

However, the malformed crop, unsuitable for market, could be harvested instead of being ploughed in, and this would amount to approximately 25 000 t/yr dry weight, or 440 TJ/yr. Overall, therefore, wastes from Brassicas only represent approximately 95 000 t/yr dry weight in accessible material, representing some 1640 TJ/yr. In national terms this amount is not very significant, and comparable results can be expected in other countries.

B6 WASTES FROM OTHER VEGETABLES

Because the mix of vegetables grown interact in a complex way with climatic conditions and cultural practices, it is necessary to study each country in detail to reach firm conclusions about vegetable waste arisings. However, French vegetable production exceeds the UK figure by a factor of 1.5; corresponding ratios for Italy and Germany are 3.7 and 0.4 respectively. Hence, if wastes were to arise in a similar ratio to marketed crop in those countries, the figures would be France, 18 000 TJ/yr, Italy 44 000 TJ/yr and Germany 5000 TJ/yr, in terms of gross energy, with some 60% of these quantities being actually collectable after modification of machines and introduction of extra harvesting steps.

Table II sets out the principal components of crop wastes estimated to arise in UK, France, West Germany and Italy.

Table II Estimated arisings of total and collectable crop residues from major crops in 4 countries of western Europe (1000s of tonnes dry matter)

Crop	UK	France	W. Germany	Italy
Cereals	12 000	28 000	21 000	8000
Maize	2	8000	500	6500
Rice	–	50	–	850
Sugar beet	1150	4000	3000	1550
Potatoes	250	350	450	120
Vegetables	700	1100	300	2600
Fruit, including grapes	30	1500	300	1760
Total	14 132	43 500	25 500	21 380
Estimated total gross thermal value (TJ/yr)	247 300	761 300	446 300	374 200
Mt oil equivalent	5.6	17.3	10.1	8.5

B7 WOODY WASTES IN AGRICULTURE

Woody wastes from agriculture comprise prunings and grubbings from orchards and vineyards. Few figures have been published on mean annual arisings per hectare, but those given by Spedding and Walsingham (1978) suggest a mean of 1.07 t/ha. This mean level of arisings from the national areas of orchards and vineyards has been included in Table II, and lead to a total of 3.54 Mt/yr, or 62 000 TJ/yr for the four countries, or 1.4 Mt of oil equivalent.

C Logistics of collection and storage: Geographical density of arisings

Unlike animal wastes, which can usually only be collected from houses and concrete yards, crop wastes require collection from extensive areas of fields; even for on-farm use, a distinct harvesting operation is required for the residues, separately from the main crop, which puts pressure on farm labour and equipment at a busy time, leads to some extra soil compaction and adds considerable costs. Even an on-farm operation using crop residues for energy has to make allowance for these costs and pressures, even when the residues have no other value and would normally be ploughed into the soil. On the other hand, utilisation off-site implies that residues must be purchased from the farmer by an outside operator and allowance must then also be made for a profit to the farmer and transportation from the farm to the energy-using plant. Moreover, crop residues are typically available only at one time of the year, although the energy demand is likely to be over a much longer period, all the year around, or most commonly, concentrated into the winter months, although residues arise mainly in the late summer and autumn. Storage is therefore another added cost, for handling into and out of the store and for servicing capital invested in the storage facility. Storage and transport costs are both related to density and physical form; therefore it has often to be decided, especially in the case of straw, whether to expend money and energy upon mechanical densification, for the sake of saving costs in transportation and storage. All these factors combine to ensure that residues arrive at the site of utilisation or conversion at a significant cost, the level of which is a factor in determining whether the residues can compete, as fuels, with fossil fuels, either at the current price, or at a projected future price which allows for fossil fuel prices to increase in real terms. Cereal straw is a good example illustrating the interactions of these factors.

McCann and Saddler (1976) studied methods and costs of straw collection in Australia; they concluded that, with specially designed large vehicles, straw could be collected and delivered to a central processing plant at a cost equivalent, in 1980 terms, to £9.60/t. An American report in the same year (Stanford Research Institute, 1976) estimated that collected agricultural wastes generally would arrive at central processing plants at prices ranging from £18.00 to £28.30/t, after conversion to 1980 terms, whereas the corresponding figures from a University of California (1976) report on rice straw were £9.25 to £11.31, more in accord with the Australian work. The truth is that a very wide range of values is possible depending mainly upon harvesting method, the physical form and density of the residue after harvesting, methods of handling into and out of transport and stores, the types of planted crop and t/ha of total land surface of the region or area concerned. These factors have been highlighted by work in Britain, France and Denmark (Plom, 1975; Morris et al., 1977; Chartier et al., 1978; Pedersen, 1979), which have taken account of the possible variations in each factor contributing to the costs. Chartier et al. studied the total procurement costs and their individual components for chopped cereal straw and maize stems. The estimated total costs for chopped cereal straw at a centralised site of utilisation varied from £37.40/t to £40.33/t in 1980 terms, including a positive purchase price to the producer. It was calculated that different levels of purchase prices resulted in different proportions of all cereal farmers being willing to sell their straw, so that high purchase price led to of relatively smaller catchment area and lower transport costs, while lower purchase prices led to fewer farmers agreeing to sell and to larger catchment areas and higher transport costs. Typically, the purchase price was about 24% of total procurement costs; the other components were gathering and loading (37%), transport (24%), storage (1%) and chopping (12%). The estimated total costs for chopped maize stems was from £32.90/t to £38.40/t, in 1980 terms, being made up of, typically, purchase price, 23%, gathering and loading, 40%, transport, 11%, storage, 1%, and chopping, 22%. Chopped straw appears, understandably, to be normally much more expensive than baled straw; for example, Pedersen found that barn-stored chopped straw supplied to an off-farm site 15 km from the farm was £32.32 in 1980 terms, but transport amounted to only 5%, with barn storage (55%) being the largest cost; farmer's profit was excluded. The same author found that big bales stored in the open provided the cheapest on-farm straw as fuel, costing the farmer only £14.55 in 1980 terms. There is a general concensus that big bales, in particular, big round bales, constitute the cheapest way of gathering and storing straw on the farm, especially on the large specialist cereal-growing farm. They are especially attractive because they conserve farm labour, which can then be devoted to more profitable activities. These big bales, however, are distinctly disadvantageous in transport by reason of both their low bulk and their shape, restricting the numbers that can be carried on trucks. The principal conclusion of Morris *et al.* (1977) was that the successful development of high density baling systems would be the key to any really widespread use of straw in centralised factories. This effect of bale density on procurement economics is well brought out in detailed costs tabulated by these authors, who considered bales varying in weight from 19 to 632 kg with densities ranging from 120 kg/m^3 up to 300 kg/m^3. Within this range three high density types of bales were considered. The type developed by the National Institute of Agricultural Engineering (Silsoe) gave the lowest procurement costs overall at transportation distances ranging from 24 to 160 km. In 1980 terms, the total procurement cost at 24 km was £7.15/t for the N.I.A.E. high density bales and £9.71/t for big round bales; at 160 km transport distance it rose to £12.05/t for the N.I.A.E. bales and £36.01 for big round bales. These authors point out that the system is relatively insensitive to the capital cost of the high density baling equipment and that development work on improved high density baling systems would be justified. The above costs do not include a profit to the farmer but, notwithstanding this they are much lower than the European estimates already cited, which is likely to be related to lower British farm labour costs. Straw may be combusted in furnaces designed to handle either the chopped form or large or small bales, so it seems likely that future centralised consumption of straw for burning or conversion, so long as it is based on combine harvesting the cereal, will be based upon an intake of high density bales.

The thermal content of straw, after allowing for the residual moisture of the air-dried material to be volatilised, amounts to approximately 15 GJ/t. Therefore, the estimated delivered cost, above, of N.I.A.E. bales at £7.15/t represents £0.48/GJ, while chopped straw, according to the Danish estimate of £32.32/t, amounts to £2.15/GJ, neither of which includes farmer's profit. This compares with heating oil, to date, at about £4.10/GJ. Hence it is understandable that some 6000 straw-burning furnaces are said to have been sold in Denmark, since the cost structure, to a farmer who has the straw available, is quite attractive.

The problems of collecting vegetable wastes arisings from the field have already been referred to (chapter 3, section B) especially in the case of Brassicas, peas and beans, mainly on account of harvesting machinery geared to streamlined operation in collecting the principal crop, or the incidence of serial hand harvesting. In the case of brassicas, not only is the residue arising quite small in energy terms, but their utility is further reduced because they tend to be not strongly concentrated in the same areas as other vegetables, often depriving them of what would otherwise have been their main role — that of a supplementary feedstock. The situation with Brassicas is also made more unfavourable because the average size of holding in an area may be quite small, necessitating that very large numbers of farms must be negotiated with for supplies and visited by transport. These combined factors of small holding size, non-coincidence of location with other vegetable feedstocks, low total arisings and low collectability, really reduce Brassicas to a negligible potential role in the provision of energy.

The cost of procurement of Brassicas and pea wastes, and other minor vegetable wastes when they arise, varies between £2.50 and £8.00/t wet weight at the farm or vegetable packing station, £15.60 to £50.00/t dry weight. Packing station waste involves no separate field collection costs and is often purchased for cattle feed at about £3/t wet weight. The cost of pea vines collected from the field would range from £3.50/t to £5.80/t wet weight, while Brussel sprout stalks harvested specially from the field would cost about £8.00/t wet weight, in the farmyard, in terms of 1980 costs. The range of costs per GJ gross thermal value is therefore £0.90 to £2.86, or after conversion to biogas with 60% efficiency, £1.50 to £4.77, for feedstock alone, processing costs and ensilement for storage being extra. The costs of procuring ensiled sugar beet tops for fodder has been estimated by Nuttall (1978) at £6.73/t, in 1980 terms at 18.3% dry matter, corresponding to £36.76/t dry matter; some economies could be made in the method of ensilement if it were done for energy purposes, especially with regard to facilities for handling silage effluent and economics of scale. However, it does appear likely that a feedstock cost of the order of £2.00/GJ of gross thermal value will prove to be general for stored anaerobic digester feedstock derived from wet green agricultural residues.

Wherever it is envisaged to use green plant waste to feed an area digester, an important parameter is the geographical density of the arisings (dry t/ha) over the whole surface of the area concerned. For example, British counties vary from Norfolk (0.65) and Lincoln (0.58) down to values well below 0.01, and only fourteen British counties exceed 0.10. This has a very significant bearing upon transportation economics, since, after incurring an initial cost for loading and unloading the truck, transportation cost varies linearly with distance (Fig. 1).

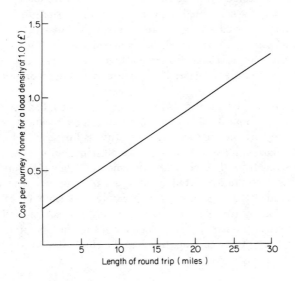

Fig. 1 Relationships between transportation costs and distance for lorry transport in full-time use. (December 1979 costs)

The geographical density of arisings of cereal straw frequently reach a level similar to the densest levels of wet crop arisings. For example, 31 departments of France produce more than 0.5t of straw per hectare of total surface. The English counties of Cambridgeshire, Norfolk, Suffolk and Essex, reach only 0.35 to 0.40, on account of both lower yields than in France and higher grain/straw ratios, but these levels, and even considerably lower ones, are nonetheless quite promising for economic collection.

Finally, attention is drawn to the concept of 'whole crop harvesting', as a revolutionary concept for cutting the crop close to the ground, collecting and carting to a central processing plant for separation into distinct components and dehydrating if necessary. There are several potential advantages of this method, the most obvious is that only one, very simple, harvesting operation is needed; however, added to this is the avoidance of field losses, so that significantly more dry matter is obtained from the crop; in the case of cereals, this applies to both the grain and the light straw components. Also with cereals there is a potentially important advantage of earlier, harvesting, which may help to avoid some losses of grain, but also frees the ground for earlier sowing of a catch crop to be grown through the late summer period, either for feed or energy purposes; such a catch crop may be segregated in the same factory as the cereal crop to yield feed fractions of different grades or distinct feed and fuel fractions. The chief disadvantage, apart from capital investment, is the need to consume fuel to dehydrate the early-harvested crop. A Swedish company who market a whole crop harvesting system, have claimed that this energy expenditure can be off-set against the energy gained by saving field losses (Vahlberg, 1978); the energy required for dehydration was said to be approximately 17% of the crop energy content, whereas the yield increase from whole crop harvesting the cereal crop was said to be of the order of 21%. These figures are in need of independent confirmation.

Work on a straw-fired furnace for use with this system has been described by Wilton (1979). The considerable potential importance of whole crop harvesting to biomass/energy programmes appears to justify its further development.

Whole crop harvesting was part of an overall scheme for collection and utilisation of rice straw from the California Sacramento Valley (University of California, 1976). The presentation detailed the types of transport needed for the purpose and concluded that the straw, which would cost from £9.25 to £11.31 to collect conventionally (in 1980 terms) would only cost £2.62 by the whole crop harvesting approach, as a result of overall savings in the operation of the farms. Even this cost, they suggested, might perhaps be eliminated in practice as a result of savings of grain losses.

D The composition of crop residues

The composition and energy contents of most of the crop residues discussed in this chapter are given in Table III. The most important observation here is that most of the residues contain significant levels of metabolisable energy, that is, energy in a chemical form which is available to be absorbed by the animal to support its growth and metabolism, and that many of the wastes contain significant crude protein levels in the dry matter. This makes them suitable, to varying extents, as animal feeds, and amounts to a competitive use to energy conversion. Indeed, all the wastes listed are potential feedstuffs with the exception of potato haulm on account of its toxicity and much of the poorer quality cereal straws, where the metabolisable energy is too low to make them worthwhile; below a metabolisable energy level of about 6 MJ/kg, the residue takes up too much

Table III Composition of crop residues on a dry matter basis, typical values

Residue	Metabolisable energy (MJ/kg)	Gross energy (MJ/kg)	Crude protein (%)	Crude fibre (%)	Crude protein digestibility (%)	Ash (%)
Wheat straw	5.6	17.6	2.9	42.2	3	7.1
Barley straw (spring)	7.3	18.0	3.8	39.4	24	5.3
Barley straw (winter)	5.8	17.8	3.7	48.8	22	6.6
Oat straw	6.8	17.9	2.8	39.4	37	5.7
Rye straw	6.3	18.2	3.6	44.7	19	3.0
Rape straw	6.5	18.0	3.0	45.0	72	4.5
Pea straw	6.5	17.9	10.5	41.0	48	7.7
Pea haulm and shucks (ensiled)	8.7	16.9	16.7	29.0	57	20.0
Potato haulm	6.5	17.3	10.9	27.0	44	13.5
Potato haulm (ensiled)	6.4	17.1	12.8	17.6	38	22.4
Sugar beet tops	9.9	15.4	12.5	10.0	70	21.2
Sugar beet tops (ensiled)	7.9	13.4	10.4	14.8	62	32.2
Cabbage (outer leaves)	11.6	16.8	18.3	11.5	–	15.0
Cauliflower	12.1	17.9	29.1	11.3	–	11.3
Brussel sprout waste	11.4	17.9	18.4	16.3	–	7.4
Bean straw	7.4	18.0	5.2	50.1	49	5.3

Compiled from data in or derived from Evans, R. E. (1960) "Rations for Livestock", M.A.F.F. Bulletin 48, HMSO, London, and "Nutrient Allowances and Composition of Feeding Stuffs for Ruminants", L.G.R. 21, M.A.F.F.

space in the alimentary canal and becomes counter-productive by excluding more nutritious food which the animal could have otherwise eaten. Indeed, the difference in metabolisable energy between spring and winter barley straw is significant in this respect and affects their usage on a major scale in practice.

The level of metabolisable energy is also somewhat indicative of the gas yields that can be expected from anaerobic digestion; for example, the work of Loll (1977) shows higher gas yields from potato haulm and beet tops than from wheat straw and, although Loll's gas yields are higher than those obtained by many workers, especially for straw, the ratio between the values bear a resemblance to the ratios between the metabolisable energy values for the same feedstocks. It appears that the animals' digestion and the anaerobic bacteria utilise much the same chemical components in the feedstocks.

Metabolisable energy is inversely correlated to the level of crude fibre in the wastes (Fig. 2).

Metabolisable energy values given in Table III have been derived from actual

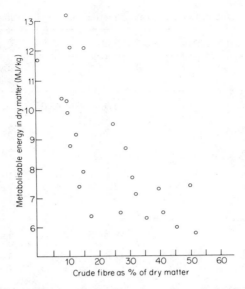

Fig. 2 Relationship between crude fibre content and metabolisable energy content in various types of biomass

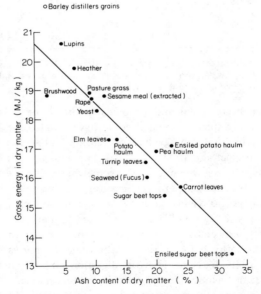

Fig. 3 Relationship of ash content and gross energy in various types of biomass. (Straight line denotes theoretical relationship extending through 20.6 MJ/kg at 0% ash and zero at 100% ash, including a 'typical' value of 17.5 MJ at 15% ash)

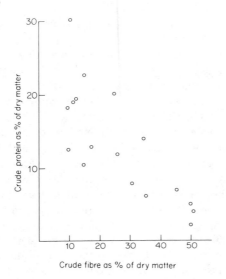

Fig. 4 Relationship between crude fibre content and crude protein content in various types of biomass

determinations of the percentage of organic matter in the residue, usually conducted using sheep, and multiplied by a factor to give the energy content of the digestible portion. The gross energy values have been calculated from chemical analysis, and not determined calorimetrically.

Gross energy values are inversely correlated with ash content, as would be expected, since ash represents weight contributing zero energy, (Fig. 3). The mean calculated ash value of all the listed residues except ensiled sugar beet tops (which seem anomalous) is 10.4; the mean gross energy level is 17.5 MJ/kg, giving a mean gross energy of ash-free organic matter of 19.5 MJ/kg. Thus, it is common to take a value of about 17.5 MJ/kg as generally applicable to typical plant matter of unknown composition and about 20 MJ/kg for ash-free plant matter.

The crude protein levels in the residues are inversely related to crude fibre levels (Fig. 4).

The compositional changes seen in potato haulm and beet tops after ensiling are of potential interest because they may reflect general experience on ensiling any green crop wastes for subsequent use as digester feedstock. It would be particularly interesting to know whether significant cellulose breakdown occurs in the silage clamp, since this would tend to augment gas production during anaerobic digestion; however, in the two examples given in Table III one crude fibre level increases and the other decreases. However, commonly ensiled feeds such as grass, mustard and clover all show increases in crude fibre during ensilement, which is suggestive of a loss of non-fibre organic matter without significant cellulolysis.

Finally, all crop wastes have a potential value as fertiliser returned to the soil, which is quite measurable in respect of the nitrogen, phosphate and potassium contents; there is also an unquantifiable benefit from the concommitant additions of carbon to the soil. A rough indication of maximum fertiliser values is given in Table IV, though crop residue contents of nitrogen, phosphate and potassium vary over a wide range. Values for most types, however, do not spread much beyond the indicated levels of £2 to £10 per tonne of dry solids. The latter figure would be a rather substantial addition to energy feedstock costs if it all had to be borne as an input charge. In practice, however, it can be diminished to take account of the non-availability of a significant fraction of the nutrients; for example, Widdowson (1974) found that the nitrogen in beet tops had little value for the following crop. Any loss of nutrients to the agricultural system would be largely avoided if the residues were used for anaerobic digestion feedstocks and the digester residues returned to the land. In this case, however, the costs of storage and spreading, which would not normally occur when the residues are burnt off or ploughed in, would have to be taken into account.

Table IV Maximum monetary and energy value of N:P:K: in agricultural residues

	Percent on nitrogen	Dry matter basis Phosphate	Potassium	1980 money value/t (£)	Energy input for equivalent inorganic fertiliser (GJ/t)
Green plant matter (typical composition)	2.4	0.3	2.5	£10.00	1.41
Cereal straw (typical composition)	0.5	0.08	0.6	£2.22	0.30

E Methods of utilising crop residues for energy

In Denmark, where straw-burning has already become quite popular, a major use is for heating farmhouses. To accomplish this all that is needed is an increased size of furnace, compared with an oil or coal burning furnace of equivalent capacity, designed to give an optimal efficiency with straw; furnaces already available for burning wood and rubbish may be quite suitable. Such a furnace can provide very low cost heat on farm, where the straw may be had for the costs only of harvesting and storage; however compared with fossil fuels, and especially fluid fossil fuels, considerable labour is incurred for stoking and the fuel is inconvenient, the furnace being unable to function for more than a few hours without attendance and being unable to self-start, and stop automatically from a pre-set time clock. One solution is to run the furnace automatically on oil when no-one is near to attend to it and stoke it with straw when possible; however, this reduces straw to the role of a supplementary fuel only. However, the engineering of convenient straw-fired central heating furnaces is a problem being actively worked upon in several centres in Europe.

Grain drying is a fuel use that would be especially well matched for straw use, since the fuel and the grain arises from the same source. Between 50 and 200 MJ/t of grain is used in the drying process, depending upon how much heat is used and how long the farmer stores his grain before selling. The seasonality of the demand makes it rather inconveniently suited to being met by anaerobic digestion, about half of the demand falling into the July–September quarter. Such an application for straw demands an efficient and reliable furnace capable of automatic stoking, just as in the case of domestic heating. Work on such a furnace for specific application to grain drying is described by Wilton (1979). Other on-farm applications are available, though only rarely will they be matched to straw production on the same farm; these included crop drying, which would require very large supplies of fuel, say, from 5 to 100 TJ/yr, and would involve collecting straw from a sizeable area. Glasshouses could be heated by straw and the requirements of the pig and poultry industry for space heating for raising young stock could be provided in this manner also, although these industries have their own manure available for anaerobic digestion.

An interesting possible application for straw is to heat anaerobic digesters, instead of using the more valuable biogas for this purpose. For example, if straw is priced at, say, £1.20/GJ and biogas is costing £4.50/GJ, in a case where the heating load to maintain digester temperature is 20% of gross output, the cost of the gas falls by 15.5%; if 40% of gross gas output would be required to heat the digester, replacement by straw at £1.20/GJ brings gas cost down by 27%. Considerable advantage may accrue in future from such substitition; it may not apply to gas cheaply produced from high concentration pig slurry and would not apply at all where gas is used for electricity generation; giving rise to waste heat, recycled to the digester; it would apply with most force to feedstocks of dairy waste or vegetable matter, where the fuel is required for sending off the plant site in gaseous form, to supply an industrial consumer, for example, or for pressurising into cylinders for vehicle propulsion.

On a larger scale straw could be used for providing process heat, space heating and steam-raising in industry, in district heating schemes and even for electricity generation at large power stations. An American report (Stanford Research Institute, 1976) evaluated the prospects for power stations fired by agricultural residues and concluded that fuel delivered to US coal-fired stations was so cheap that residues could not compete, whereas, at that time residues could just compete at oil-fired stations.

Straws, and all reasonably dry residues, are amenable to thermal conversion processes for the production of gas or oil (see chapter 2). An evaluation of the production of pyrolytic oil from straw is given by McCann and Saddler (1976); work on straw gasification is in progress by several equipment manufacturers (Lucas, 1978).

A partial thermal conversion (pyrolysis) process may be used to effect a compression and forming of the material to give dense, regularly shaped blocks or briquettes. (Hansford, 1974). It is claimed that, with dry straw, only 5% of the thermal value is consumed in the pyrolytic stage, 12% for straw of 50% moisture content. Conversion to a convenient solid fuel might also be accomplished by variations of some of the processes now being developed for producing powdered or pelletised fuels from municipal refuse.

Utilisation of wet green material, such as vegetable waste, will generally necessitate anaerobic digestion: a biological approach is needed to separate a fuel component from the wet biomass without the energy-intensive step of dehydration. The optimum mechanical arrangements for handling plant waste through a digester on a commercial scale have yet to be developed; it seems inevitable that some form of chopping or pulverising will be necessary either at the digester site or at the point of harvesting. Quite large digesters should prove practicable in vegetable-growing areas with arising-densities like those of the eastern counties of England; for example a density of 0.4 dry t/ha/yr permits a 50 dry t/day plant to be fed from a circular catchment area radius 12 km, mean journey distance 8.5 km.

Alternatives to digestion are few for this class of feedstock. The composition is generally not suited to alcohol production from the whole material on account of the high cellulose and hemicellulose content. Air-dried material could be combusted or thermally converted, like straw, but opportunity for effective air-drying occurs only rarely in northern Europe.

One future perspective of note for green plant matter is crop fractionation, which would yield protein feed concentrates and solid fuel as well as a sugary solution suited either to low-residence-time digestion or, possibly alcohol production. This is discussed further in chapter 5.

F Competitive uses of potential crop waste as energy feedstocks

F1 GENERAL

All agricultural residues have to be disposed of if they are not used, yet most disposal methods, such as rotting and burning, have useful effects; rotting vegetable matter adds both humus and nutrients to soil; burning adds nutrients to the soil, helps to prepare the ground for the next crop and may dispose of pathogens. Disposal and use are therefore closely related, since there are few residues arising in agriculture today of which absolutely no use is being made. In making a claim upon them for energy purposes, one has to show that benefit will accrue to the farm and profit to the farmer by following this course, rather than any other.

F2 STRAW AND OTHER DRY RESIDUES
(a) *Agricultural uses*

In the case of straw, Europe's principal crop residue, numerous competitive uses exist which are a distinct advantage to agriculture and, in some cases, to industry. The net effect is that the quantities of straw potentially available for energy do not approach the total levels of arisings given in Table II. Other residues, where they are collectable, are generally more free from competitive use. Therefore, although in the European scene straw predominates quantitatively among residues that are potentially available for energy, the degree of predominance is not so great as Table II would suggest. For example, Palz and Chartier, 1980, estimate that approximately 25% of all straw arisings is as much as could be employed for energy within the nine countries of the European Communities without disturbing important present uses in agriculture and elsewhere. Probably as much as 60% is utilised for animal bedding, although the proportion varies considerably from country to country. A part of this (about 10% of the bedding) is usually reckoned to be eaten by the animals and the remainder becomes a component of livestock industry waste. This usage is expected to slowly decline as modern loose or cubicle housing methods, requiring zero or minimal bedding, spread through the dairy industry. However, this is a gradual process and for the forseeable future bedding can be expected to remain by far the biggest single use for straw. Some areas have much more unutilised straw than others; for example, Denmark burns about 36% of all straw, while in France, the Centre-France regions burns 40% of the total compared with a national average of 12%. Bavaria is outstanding for having few other uses for straw, since almost 90% of all harvested straw is utilised for animal bedding.

Straw is also used as feed for animals when the quality is high enough; as can be seen from chapter 3, section D, it is the metabolisable energy content (ME) which chiefly determines the quality for this purpose. Barley straw, especially straw from spring barley, is mainly suited to feed use and in the UK some 43% of all harvested barley straw is fed, compared with only 4% of harvested wheat straw, at least 70% of harvested wheat straw being used for bedding and crop protection.

Other traditional straw-using industries exist which, though important in themselves, make only very small demands upon straw supply; these include thatching and the mushroom industry.

The demand for straw as feed may well increase in future and take enough straw to seriously compete with energy use. Since the role of straw in feed is as a low cost source of metabolisable energy, any process that increases the proportion of the straw gross energy which is metabolisable would be economically interesting. This is especially true because at the levels of metabolisable energy in straw, some 5 to 7 MJ/kg, the balance of advantage to the farmer is very fine; 5 MJ/kg is really too low to be useful; all feed takes up space in the gut and depresses the appetite for other feeds; if the energy content is too low, the feed is simply not worth its place in the diet. A value of 7 MJ/kg is advantageous, however. Processes for treating straw with alkali or ammonia have been developed (Robb and Evans, 1976; Greenhalgh and Pirie, 1977; Arnason and Magne Mo., 1977; Wilkinson, 1979; Bowerman, 1977; Wilson and Brigstocke, 1977) which, in one simple step, increase the metabolisable energy level in lower grade straw up to about 8.6 MJ/kg. Initial ventures to apply this process on a commercial scale have had fluctuating fortunes as the prices of other feedstuffs have gone up and down, but the long-term future of the processes may well be quite good. Present usage of alkali-treated straw has been mainly in compounded feedstuffs, into which it can be introduced at about 10%. At present, in most of NW Europe, this would restrict the market to a level at which straw consumption for the purpose would be only some 3% of arisings, or 6% if compounded feed usage were to double in future. However, development of safe on-farm methods of alkali-treatment, especially ammoniation, which does not involve the same problems of residual alkalinity and increased ash content as treatment with sodium hydroxide, would open the possibility of including the treated straw at a 10% level into all cattle feed; even if its use were confined to the winter period, the use of straw for feed might well then claim up to an additional 10 to 20% of total straw arisings and wipe out most of the straw surplus that is otherwise available for energy purposes. At the present time it is impossible to see whether this will happen; probably it will not happen for a considerable time for, although ammoniation is a simple process, there is usually much inertia and reluctance encountered when any suggestion is made to introduce processing operations onto farms, with only a very small percentage of farmers being sufficiently technically minded to accept the challenge.

(b) *Non-agricultural uses*

Dry agricultural residues are amenable to quite a wide range of industrial applications, including (*i*) particle board manufacture, (*ii*) paper-making and (*iii*) feedstock to chemical and biochemical industry. In a sense the first two of these may be bracketed together, since they both use the fibrous structural properties of the straw. Italy has a high usage of straw for industrial purposes, amounting to 18–19% of the national arisings. (Triolo, 1977) mainly for paper and board. Holland, W. Germany and Romania all operate plants to utilise straw cellulose for paper-making, Romania having the largest plant for this purpose, with an annual capacity of 200 000 tonnes (Rexen, 1976). Where wood pulp is available in sufficient quantities the paper and board industry has always preferred to use it, so future trends in usage of straw for this purpose will depend upon the balance of supply and demand in the international trade in wood pulp and upon the industry's adaptation to the use of straw. Adaptation certainly is needed; problems arise from the high ash content of straw, 5–10% for cereals, up to even 15% for rice straw, which causes scaling on evaporator boilers and other parts of the system, reducing thermal efficiency. Furthermore, straw pulp is more difficult to wash than wood pulp, consuming more water and forming more effluent (Truman, 1974). However, it is clear that these problems can be overcome and that reasonably good papers can be produced, containing from 20–60% of straw, and this would greatly ease the shortfall in world wood pulp supplies. Projections show (Truman, 1974; Palz and Chartier, 1980) that maximum likely usage of straw in paper and board making in the UK and in most W. Europe situations, would amount to some 15% of straw production. It is clear from this that animal feed use and paper making use have the potential together to wipe out the straw surplus, but there is no way of forecasting whether they will actually do so.

Paper and board making can be of interest using other crop residues, including sugar cane bagasse and cotton linters (Rexen, 1976). Bamboo can also be employed and, while this is not a residue, it does serve to indicate that a range of different types of biomass can be employed.

Straw can also be projected as a potential feedstock to chemical and biochemical industry, to help substitute for oil-based chemicals, though such use has not yet been established. The cellulose component may be hydrolysed to yield glucose, from which ethanol, citric acid, butanol, synthetic rubber, ethylene, polythene, acetone, polyvinyl-chloride (PVC), sorbitol, lactic acid, antibiotics, yeast and other single cell protein, surfactants, textile chemicals, pharmaceutics and pesticides may all be derived by chemical or microbiological transformation. The initial hydrolysis step, in the versions that are most practicable today, call for a high temperature, and therefore energy-intensive, treatment with mineral acid. Hemicelluloses, also present in the straw yield pentoses, especially xylose, upon hydrolysis and can be converted to xylitol, furfural, furan resins or furfuryl alcohol. The lignin fraction of straw is a potential source of phenolic chemicals, such as vanillin and catechols. Various versions of the acid hydrolysis step have been published by Porteus (1976), Converse *et al.* (1971) and Sachetto (1977). One version (Hansford, 1974), would entail recovery of the unhydrolysed fibrous portions, so that both chemical feedstock and particle board would result from the process. Another promising development (Rijkens, 1977), uses mild hydrolysis, to attack only the hemicellulose, leaving the cellulose fibre intact for paper making. The quantitiative scope for absorption of straw by the chemical and biochemical industries cannot be foreseen. In the long term it seems inevitable that biomass and coal, between them, must substitute for all petrochemical feedstocks, but this general observation leaves the size of the future role of straw in this speculative area quite uncertain, though potentially huge, at a level where it would inevitably be limited by supply.

F3 COMPETITIVE USES FOR WET AGRICULTURAL RESIDUES

The main competitive use for green plant matter is for animal feed. That many of the residues are not utilised for this purpose is attributable in many instances to (*i*) relative lack of suitable livestock in the locality, (*ii*) the time of arising coinciding with adequate availability of fresh forage from other sources, (*iii*) lack of facilities, machinery or labour for ensilement and (*iv*) lack of contact between livestock farmers needing feed and arable farmers with the residues. There is no doubt that beet tops and most vegetable-growing residues are excellent feeds (e.g. Nuttall, 1978). It is equally certain that ensilement is needed to make the fullest use of them.

In the UK, according to Nuttall, approximately 25% of sugar beet tops are utilised for animal feed, of which 18% is fed in the field and 7% is carted off the field to be used as fresh feed or silage. In future, especially with the spread of the practice of silage-making, more use may be made of these materials, but especially of sugar beet tops.

If crop fractionation is practised for grass and other fodder crops and, especially, if catch crops or plantation energy crops are subjected to crop fractionation as suggested in chapter 5, then a similar opportunity would be opened up to fractionate crop wastes in the same processing plants and hence utilise different components of the crop for different purposes. In such a case, the green crops and residues would be dejuiced and a fibrous solid material would remain, which could either become low to intermediate grade ruminant feed, solid fuel, or feedstock for production of chemicals.

[4]
Animal wastes

Animal excreta and bedding are a major potential source of materials in parts of the world where agriculture is orientated towards animal products, or where animal power, from bullocks or horses, is employed on the land. Beasts of burden similarly contribute. However, as fuels these wastes have inherent disadvantages. They are wet, often foul smelling and unstable. They represent fuel at all only in so far as the animal producing them is an inefficient utiliser of its food. Forage, grain or other feeds given to the animals are digested and assimilated as completely as possible, including the energy-yielding carbon compounds which are also potential fuels. It follows, therefore, that animal wastes are relatively enriched in those materials, such as ash, which have no energy value or in those materials such as lignin-coated fibres, where the energy value which is present is in a metabolically unavailable form. It is helpful to distinguish, therefore, between metabolisable energy, which can be used by farm animals, and total thermal content, which includes the unmetabolisable fractions. When metabolisable energy or protein is present in sufficient amounts in animal wastes they are sometimes re-incorporated into diets (Wilkinson, 1978). The most appropriate materials to use as biomass fuels are often those that contain an insufficient proposition of their gross energy in metabolisable form to render them of interest as agricultural feedstuffs. The normal fate of such materials is to add humus to the soil or to be burned to waste.

The main routes available for utilising animal wastes as fuel at present are (*i*) direct combustion after sun-drying and (*ii*) anaerobic digestion to yield biogas, a converted fuel which is devoid of the disadvantages inherent in the animal wastes from which it is prepared. The sun-drying route is only available in areas where the climate is dry and warm for long periods.

A Characteristics of animal wastes

Table I gives figures for typical composition of waste from farm animals, though in practice values can fluctuate according to variations in diet and methods of collection. The whole comprises faeces plus urine; as the faeces are a residue left

Table I Typical composition of total wastes from fully grown farm livestock. (Faeces and urine)

	Hens (2 kg)	Pigs (50 kg)	Beef cattle (450 kg)	Dairy cattle (500 kg)*
Total solids (dry weight, kg/day)	0.026	0.315	3.48	4.92
Volatile solids (% dry basis)	70.0	85.0	80.0	80.0
Nitrogen (% dry basis)	5.0	4.5	3.7	5.0
P_2O_5 (% dry basis)	4.8	2.7	1.1	2.0
K_2O (% dry basis)	2.0	4.3	3.0	5.0

* UK data; other data from Taiganides and Hazen (1966).

over from digestive processes in the gut they contain any undigestible components of the food. This is always likely to include, even in ruminants, which digest cellulose, fibres of ligno-cellulose derived from grass or straw, where lignin around the cellulose fibres prevents access by digestive enzymes. Similarly, in the case of animals fed on grains, the husk or shell may sometimes pass through relatively unaltered. In addition, the faeces contain the bodies of the bacteria of the gut flora, comprising complex bacterial cell wall constituents together with cell contents such as protein and nucleic acids. Lastly, the gut contents are continuously treated, during their passage through the gut, with various enzymatic secretions and mucuses together with one secretion, the bile, which has both a digestive and an excretory role. Therefore faeces, containing residues of all these materials, are generally coloured by the bile pigments, such as bilirubin and biliverdin or their decomposition products. The urine, on the other hand, is formed in the kidney by a simple process of filtration of blood followed by selective reabsorbiton of non-waste components and concentration of the remainder. Therefore, soluble blood products, particularly nitrogeneous wastes, and inorganic salts form principal components of the urine. In man and most mammals the principal nitrogenous excretory product in urine is urea, while in birds, including the chicken, it is uric acid. In spite of their composition animal wastes do not differ widely in thermal content from that of organic matter in general, and estimates range from about 14 to 22 MJ/kg compared with the typical value for plant material of about 17.5 MJ/kg on a dry basis. These values are boosted by the nitrogenous excretory products and often by lignin (25 MJ/kg), but diminished by high ash contents.

Although diet affects directly the composition of the wastes, those from certain species nonetheless retain certain features. The most important distinction is that between the cellulose digesting ruminant, such as sheep, cows, buffaloes, horses, elephants etc. and non-ruminant species. The ruminant does not produce cellulose digesting enzymes itself but these are produced by the rumen bacteria to hydrolyse cellulose to simple products that can be assimilated. Ruminants tend to have a diet which is high in roughage and hence the resulting faeces, containing the undigestible portions of the fibre, are normally much higher in fibre than in the case of non-ruminants. The very active microbial population in the rumen ensures that most accessible metabolites in the food are utilised; the faeces from ruminants are therefore low in substances that support the metabolism of further bacterial species in a digester and the yield of fuel gas from cattle waste is only moderate (e.g. 200–250 l/kg of solids, CH_4 content, about 55% by volume). Wastes from pigs and chickens generally show much better production of fuel gas, since most of the cellulose component of the food remains in the waste and can be utilised by the methane-producing microorganism populations (e.g. 400–500 l/kg of solids, CH_4 content, about 60–70% by volume).

There are considerable effects from the use of straw, wood shavings, etc. as animal bedding. These become intimately mixed with the animal wastes and in this case the contribution to thermal value from bedding may be comparable, on a dry matter basis, with the contribution from the animal waste itself. As the bedding materials are largely comprised of ligno-cellulose they are relatively unamenable to digestion by microbial activity and, if they are to be used as fuel at all, lend themselves best to direct combustion. This may cause no problems if it is planned to simply sun-dry faeces and burn them, but straw and wood shavings can cause a variety of problems in handling in the more sophisticated systems of anaerobic digestion and, indeed, may require to be removed entirely from the system to prevent it from becoming clogged and inoperable.

A1 AVAILABILITY OF ANIMAL WASTES

When sheep or cattle roam free the wastes merely fertilise the pastures, and the farmer has no practical option to employ this material differently unless labour is cheap enough to enable it to be collected. This is most likely to be done in hot dry situations, where the waste has spontaneously sun-dried and the whole thermal content of 18 ± 4 MJ/kg may be released by combustion. On the other hand, animals that are housed all the time drop all their wastes in the house and, where there is a hard smooth floor, these may be scrapped to a store to await utilisation. Where animals are housed for part of the year only or, sometimes, only at night, their waste is partially recoverable. Other practices may also affect availability, such as the assembly of dairy cattle in collecting yards prior to milking; wastes dropped during milking are also collectable. These amounts, however, are far less than the total production of animal wastes, particularly on a worldwide basis, since housing depends upon the employment of substantial capital in agriculture and upon the climatic need for such housing. Housing is associated with cooler, wetter climates; wastes from housed animals are less inclined to spontaneously dehydrate, so that they are potential feedstocks to anaerobic digestion, for recovery of only some 30 to 60% of their gross thermal value as gas, but, on the other hand, their collection may be associated with a low labour requirement, or can be automated. In the UK, the winter housing of dairy cattle varies from 105 to 180 days per year depending on location, and total collectability, taking into account the small amount of collectable summer arisings, varies from 35 to 60% depending on location.

Apart from limited collectability, other restrictions may also apply to the use of the wastes as fuel. They relate mainly to the 'developed' agricultural situation, where utilisation as fuel would be by means of process plant. Firstly, the size of animal groups or herds, may be a constraint. These affect the quantity of waste arising in one location and hence determine whether it is justified economically to install a digester or arrange transport to the nearest digester. The position is also affected by seasonality.

Dairy cattle, for example, which are housed for 6 months of the year or less, give rise to little collectable waste during the summer months and a digestion plant may be under-utilised or unutilised during that period, affecting economic viability of the enterprise on account of the continuing capital charges for the digester. The precise seasonal distribution of arisings depends on housing and husbandry practices (Fig. 1). Where farms are not large enough to justify their own digestion plant, distance, or the geographical distribution of animal populations, become signficant. Farms too small to carry out the digestion on site,

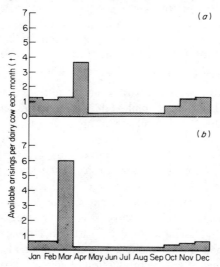

Fig. 1 Seasonality of arisings of available dairy waste as affected by housing practice. (*a*) an area with the largest contribution in the form of cubicle housing, (*b*) an area with the largest contribution as loose yard housing. (The areas also have a difference of latitude which affects the month in which the peak occurs)

ANIMAL WASTES

must be near enough to other farms to share a digester for the economics to be attractive.

Another factor is dilution, arising through the use of wash water in animal houses and through the access of rain to manure stores. This is a common problem with pig manure, where a very dilute slurry may sometimes result from the use of liquid feeds, such as whey and the too liberal use of wash water. Anaerobic digestion cannot be justified when the concentration falls too low because it would take more fuel to heat the slurry to digester temperature than would be gained from the digestion process. In this case the animal waste has become virtually 'unavailable' as fuel unless its treatment is essential for reasons of pollution abatement.

The pattern of availability of animal wastes for fuel use has been a chance event, arising from the present pattern of agriculture. It would be possible to reorganise agricultural activities so as to attach a greater priority to the availability to livestock waste as feedstock for fuel production. This would involve keeping more of the ruminant animals in houses or yards and feeding by bringing the food to the animals rather than grazing (zero grazing). This would have several advantages, especially in preventing the 'poaching' of grasslands by the hooves of grazing animals during wet weather and in terms of increased yields of grass. On the other hand, the general health and vitality of cattle housed 100% of the time sometimes suffers.

Waste from sheep anywhere or goats in the Mediterranean area, is generally considered unavailable and, indeed, quite inaccessible since housing of these animals is not the usual practice.

A2 STORAGE OF ANIMAL WASTES

Present systems of on-farm storage involve slurry tanks for diluted pig and cattle waste, and walled stores, or stores bounded by an earth mound for the more solid wastes, including farmyard manure containing straw. Investigations, particularly with manure heaps, have shown that most on-farm methods of storage are grossly inefficient in that they allow access of air to the stored material, so that there is a loss of carbon through aerobic oxidation; other losses, especially of nitrogen may occur through run-off of a liquid fraction and leaching by rain water. Precautions against these processes are necessary if manure is to be stored ready for fuel use and, perhaps the most important measures are the introduction of a cover to protect from rainfall, minimising the surface-to-volume ratio of the stored material and possibly covering its surface.

The characteristics of animal wastes are affected by the conditions of collection and storage, for example wastes from battery hens are accummulated in pits below the cages in a well-ventilated environment and tend to dry to about 70% solids content: cattle slurry stored in tanks for several months undergoes an increase in ammonia nitrogen and a decrease in total nitrogen.

A3 QUANTITIES OF ANIMAL WASTES AND GAS PRODUCTION POTENTIAL

To illustrate the potential contribution of animal wastes to energy supply in a temperate climate, Table II gives details of waste arisings from farm animals in the countries of the European Community. The estimated quantities of waste from the individual classes of farm animal are given on a dry matter basis: since the thermal content resides solely in the organic component, the gross energy content has been estimated, after making an allowance for the inorganic (ash) component. The estimate given for gross energy content is that which would be released if the wastes were dry and suited to combustion. The lower figures, in the last two columns of the table, indicate the maximum potential production of gas by anaerobic digestion, though they do not take account of the energy requirement for heating the digester. When the gas is applied to a use, such as electricity generation, which yields low-grade heat in a form suited to digester heating, the whole of the biogas is available; where this does not apply, a proportion of the gross gas output must be sacrificed for digester heating. The figures given apply to the case in which all the waste produced is collectable; since a large proportion is known to be dropped on the fields, the actual realisable biogas production, while agriculture continues to be conducted as at present, will be considerably less than shown. For most of the countries listed, the maximum potential production of biogas from animal wastes is close to 1% of the national energy requirement, and falls, at least to 0.5%, when availability factors are taken into account.

Table II Estimated national arisings of poultry, pig and cattle wastes: EEC countries gross energy contents and the energy theoretically derivable as biogas. (Bedding excluded)

	Waste arisings (1000s of t/yr of dry matter)				Estimated gross energy content (TJ/yr)*	Probable maximum[†] gross biogas production potential if the waste were all available	
	Poultry	Pigs	Cattle	Total		TJ/yr	Mtoe/yr
Belgium and Luxembourg	156	674	3357	4187	75086	29886	0.68
Denmark	147	1131	3641	4919	82335	35493	0.81
France	1293	1481	26334	29108	486295	180666	4.11
Germany	1039	2919	17840	21798	364025	146477	3.33
Holland	486	1150	5953	7589	126466	52359	1.19
Ireland	61	136	6430	6627	111206	38867	0.88
Italy	1309	1286	10551	13146	218108	88589	2.01
United Kingdom	973	1055	14435	16463	273212	114079	2.59

* At 21 GJ/tonne of dry organics
[†] Assuming 7.125 GJ/t of dry organics for cattle waste, 14.625 GJ/t of dry organics for poultry and pig waste

ANIMAL WASTES

Calculations of waste arisings and potential energy production are often most conveniently related to animal numbers, particularly when calculating in respect of individual farms, or to area of land, when considering the economics of transporting waste to a central plant for processing.

Waste from a dairy cow may be taken as about 41 kg/day, which, at a dry matter content of 12%, amounts to some 4.9 kg of dry matter per day, containing 82 MJ/day of gross energy, of which a maximum of 28 MJ/day may be realisable as biogas. However, dairy herds comprise many young animals and cows in calf as well as milking cows. The make-up of an average British dairy herd is given in Table III. The estimated weight of wet waste of arising from dairy cows is 15 t/yr but the large number of smaller animals in the herd brings this down to an average of 10.8 t/yr for the herd as a whole. These figures will vary somewhat from country to country depending upon the breeds of animals used and the make-up of the animal populations. Table III also gives the categories of pigs making up the Danish population (Green Europe, 1978) and shows that although 2.74 t/year of wet waste may be obtained from each breeding sow in pig, the large number of much smaller pigs in the population brings the average per animal down to 1.04 t/year. Further, although a laying hen produces an estimated 0.055 t/yr of wet waste, the large numbers of breeders, broilers and pullets reduces the overall mean for the chicken population to 0.029 t/yr/bird. These figures are convertible into gross energy and potential biogas production as shown in Table IV.

Table III Arisings in relation to animal population, age and type

DAIRY HERDS (UK)	Percent of dairy herd	Estimated arisings/animal (wet wt, t/yr)	percent of waste produced
Dairy cows and heifers in milk	47.0	15.0	65.3
Dairy cows in calf	5.0	15.0	7.0
Heifers in calf	10.7	9.0	8.9
Other cattle for dairy herd replacements, over 2 yrs	2.3	10.2	2.2
1–2 yrs	11.8	7.7	8.4
Other cattle, 6 mth–1 yr old	11.5	5.1	5.4
Calves, under 6 mths	11.0	2.2	2.2
Bulls for service, over 2 yrs	0.5	10.2	0.5
Younger bulls	0.2	7.7	0.1
	Total 100.0	Mean 10.8	Total 100.0

Table III (Continued)

PIGS (DENMARK)		Percentage of population	Estimated arisings/ animal (wet wt, t/yr)		percent of waste produced
Breeding boars		0.4	1.79		0.7
Breeding sows in pig		7.3	2.74		19.1
other sows and old boars		6.4	1.79		7.5
Young breeding sows Over 50 kg		1.7	1.61		2.6
Piglets under 20 kg		34.4	0.37		12.0
Other pigs 20–50 kg		27.0	0.69		17.9
Over 50–80 kg		20.6	1.61		31.7
Over 80 kg		4.2	2.12		8.5
Sow and litter		–	5.44		–
	Total	*100.0*	*Mean*	*1.04*	*Total* *100.0*
HENS (UK)					
Breeders		4.5	0.035		5.3
Layers		36.6	0.055		68.4
Pullets		13.4	0.017		7.7
Broilers		45.5	0.012		18.6
	Total	*100.0*	*Mean*	*0.029*	*Total* *100.0*

B Animal bedding

Large quantities of dry residues, especially straw, are used in bedding farm livestock and become mixed with excreta, adding to the bulk of waste which is potentially available from livestock enterprises. For dairy cattle in bedded courts and cowsheds some 5 kg/day/animal of straw may often be used, range 3–15 kg, which approximates to the dry weight of excreta produced by the animals, and the resulting manure is about 50% straw, 50% excreta solids (Robertson, 1977). Dairy cattle in cubicle housing with solid floors are bedded with an average of about 1.5 kg/day of straw, whereas cubicle houses with slatted floors require no bedding. Pig waste most commonly arises as a slurry, without bedding. The broiler side of the poultry industry uses bedding of chopped straw or wood shavings, although the quantity is not large. On the other hand, straw usage for cattle bedding is a major use, and has significant impact upon the potential availability of straw for energy and other purposes (Ader and Buck, 1979) and in the EEC as a whole it seems likely that about 40% of the available straw is used for bedding (Palz and Chartier, 1980) about 30 Mt/yr, a quantity probably of the same order of magnitude as that proportion of the 108 Mt/yr total livestock excreta which arises from housed animals. In the UK straw comprises an estimated 22% of the total livestock excreta arisings, but a far higher percentage of

ANIMAL WASTES

Table IV Types of animal; typical waste outputs and potential production of biogas. (Bedding excluded)

Type of animal or activity	Waste output t/yr (wet wt)	t/yr (dry wt)	Estimated output of gross energy (GJ/yr)	Estimated maximum potential gross biogas production (GJ/yr)
Dairy cow in milk or in calf	15.0	1.8	30.8	10.2
Average dairy herd member (UK)	10.8	1.3	22.2	7.4
Dairy cow in milk or in calf plus average no. of followers (UK)	23.0	2.8	47.3	15.7
Breeding sow in pig	2.74	0.27	4.69	3.21
Average member of pig population (Denmark)	1.04	0.10	1.78	1.22
Mature hen	0.0330	0.0095	0.142	0.097
Laying hen	0.0550	0.0159	0.239	0.163
1000 layers and associated immature birds (UK)	50.7	14.7	220.4	150.6
10 t of poultry meat produced (UK)	15.02	4.36	65.27	44.61

collectable arisings, while the other EEC countries, on average, use straw more freely.

These straw usage practices directly affect the form in which the livestock wastes arise, those with a high straw component being solid. This has a bearing upon the feeding of such residues to anaerobic digesters; while digesters designs exist for handling solid animal wastes, they are labour intensive and gas production is usually slow (Isman, 1978). Feeding to a high-rate slurry digester entails first slurring the solid farmyard manures and chopping the straw, and keeping the solids well mixed within the digester to prevent crust formation. The use of straw therefore increases substantially the amounts of livestock waste arisings but complicates their handling and utilisation for energy purposes.

C Sizes of livestock unit justifying a digester installation

Farmers and those involved in agriculture advisory services commonly need some idea of the minimum size of livestock unit justifying its own on-farm digester. Unfortunately, there is no simple answer to this question, as it depends upon many circumstances such as the gas yield from the particular waste, whether the labour involved is to be treated as a cost or as an uncosted 'fixed' overhead, and the particular use to be made of the gas. The economic outcome is also dependent upon the value placed upon the feedstock and digester residues and upon the distribution of energy use in time, since a digester produces more available fuel in summer than in winter and produces it at a more or less constant rate from day to day; moreover, storage facilities for gas are expensive installations. Figure 2 shows the calculated variation in gas costs with digester size for dairy cattle slurry and pig slurry based upon assumptions that feedstock and residues are of equal value as fertiliser and that uses are available for the gas, exactly matching its availability, including the summer/winter variations. This would apply in practice if the digester were undersized for the given gas use, the difference being made up by fossil fuel. It is assumed that the farmer incurs operating costs amounting to £5/m^3 of digester capacity; this is an arbitrary way of acknowledging that the farmer inevitably incurs some significant cost in running a digester and in handling materials into and out of it, even then these defy quantification in precise cost-accountancy terms.

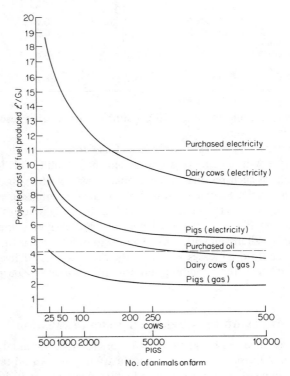

Fig. 2 Relationship of projected fuel cost to livestock farm size for gas or electricity produced from dairy or pig waste

It has been further assumed that the feedstock enters the digester at a solids content of 10% in the case of the dairy slurry and 5% in the case of the pig slurry. Finally, both groups of animals are assumed to be housed all the year round (a relatively uncommon situation with dairy cows). Costs are presented as a function of the size of livestock unit both for gas generated from the two sources and for electricity generated using the gas. For both pig slurry and cattle slurry the costs of production rise steeply with decreasing size of livestock unit, owing to the relatively high capital costs of small-scale digesters. However, gas produced from pig slurry, at the 5% solids level assumed, is always cheaper than gas from a dairy unit of comparable size, on account of the higher gas output and shorter residence time in the digester. These costs are naturally reflected in the cost of electricity generated from the gas. The costs of oil and electricity (at the farm tariff) are entered on the same figure; these serve to show that, except for the very smallest sizes of unit, pig farmers have little difficulty in producing gas at a cost that competes with oil for heating or drying. Similarly, electricity produced from such gas will be well below the cost of electricity from the mains supply. With dairy waste, on the other hand, the cost of gas for heating only begins to compete with oil at the largest sizes of livestock units, and even here the cost advantage is so slight that oil is still likely to be selected on account of convenience and easy storage. Similarly, electricity produced from biogas from dairy wastes is always close to or greater than the price for mains electricity; the margin, even for a 500 cow herd kept under the conditions assumed, is so slight that almost an 80% utilisation of the generated power is needed to break even with the price of electricity conveniently purchased from the mains supply at the time when it is needed. In many ways the estimated costs given represent the best situation that is ever likely to arise in practice; should the feedstock be regarded as having a cost, or if the solids content of the feedstock should fall below the levels assumed, or the gross gas yield fall below the 5.7 GJ/dry tonne assumed for dairy slurry or 11.0 GJ/dry tonne assumed for pig slurry, a deteriorated economic prospect would prevail. These figures explain quite well why it is that few farms yet process dairy waste on its own for biogas production, while anaerobic digestion of pig waste is economically attractive and gaining steadily in popularity among farmers.

Assuming that all the gas or electricity can be used as it is generated, the data in Fig. 2, precisely interpreted, mean that almost any size of pig farm can consider a digester for producing gas for heating and, indeed, for electricity generation as well, subject only to the availability of sufficiently small generating sets. On the other hand, a dairy unit with about 300 cows, housed all the year, can reasonably consider producing biogas for heating purposes, while a unit of 200 cows may just be able to produce electricity at a competitive cost, depending upon local tariffs.

C1 CATTLE HERD SIZES

It is apparent that in the case of cattle units, but not in the case of pig units run under conditions approaching those assumed, economic viability is very dependent on size. Countries differ markedly in average herd sizes for cattle and in the percentage of all cattle on holdings of 100 or more. In most countries, most cattle still tend to be in quite small herds, below 100 head; for example, Germany in 1978 had only just over 5% in holdings above 100 head; the corresponding figures for Italy and France were 10.7% and 13% respectively. Denmark (19.3%), Holland (22.9%), and Eire (24.3%) are somewhat better placed for using anaerobic digestion, whereas in the UK 42.2% of all cattle were, in 1978, in units of 100 head or more (5% in holdings of over 500 cattle in 1977). These figures illustrate the extent to which herd size limits the practical availability of energy from cattle wastes so long as exploitation is restricted to the on-farm situation. Wastes from the smaller farms are likely to be solid, being from animals bedded in straw, and these lend themselves towards collection for processing at neighbourhood or area plants. However, in most countries there is a continuing trend towards increasing size of livestock unit, so that the applicability of on-farm digesters can be expected to be rather more widespread in future.

D Competitive uses for livestock wastes

The commonest fate of livestock wastes on farms, whether it be viewed as 'utilisation' or 'disposal', is spreading onto land. It can often be shown that, especially where agitated above-ground metal slurry tanks are built for storage, the costs of collection, storage and spreading often exceed the fertiliser value of the material, rendering it a liability rather than an asset. However, something must always be done with these wastes. The farmer who has land suitable for spreading will tend to use it, whereas the intensive livestock farmer with little land is often in difficulties. Although, no doubt, in many farming systems, the return of the nutrients of livestock wastes to the soil is important, rather than using chemical fertilisers, the processing of livestock wastes via anaerobic digestion is completely compatible with returning the nitrogen, phosphate and potassium of the original wastes back to the land in the form of digester residues. In this case, fertiliser use is not competitive with energy use, but complementary to it. The mushroom industry is a user of animal waste in preparing mushroom compost, but only small quantities are involved, with horse manure being most preferred.

The relative absence of competitive use is reflected in numerous publications dealing with methods for destruction of livestock waste (e.g. Loehr *et al.*, 1973) and removing the nutrients most likely to cause trouble after discharge to water courses. This illustrates that livestock wastes could possibly become sources from which nitrogen (as ammonia), phosphate and potash could be extracted for use as separated fertilisers; however, a lack of practical steps in this direction presumably indicates that, at present, this cannot be economically accomplished. However, Yeck et al. (1975) stated, in a review of options for animal waste utilisation, that: 'of all the concepts for utilisation of animal wastes, the recovery of nutrients for feeding to animals appears to be the highest value use for which a practicable technology exists'. This is undoubtedly true and Wilkinson (1978) has reviewed the many experiments conducted to demonstrate the feed value of unfractionated wastes. The work has demonstrated that excreta may be included in feeds without serious harm but there are inevitably problems from the low metabolisable energy levels even though some, such as poultry waste, may have a useful content of quite digestible proteins. There is a tendancy for animals to respond by increasing dry matter intake, but even this often does not serve to prevent losses in live-weight gain or milk production. These results suggest that really widespread use of excreta nutrients in feeds must await techniques of fractionation, so that the valuable proteins may be separated from ingredients which seem to fill the animals' stomach without contributing nutritionally. Feeding whole excreta would be competitive with energy use but the feeding of fractionated excreta nutrients, like fertiliser use, would be complementary to it. Microbial protein present in the excreta, or formed during anaerobic digestion, may be separated from the residues after digestion of dairy wastes in forms containing from 27 to 56% of crude protein. (Plaskett, 1980a, b). Hence, it may be envisaged to produce both methane gas as fuel, and animal feed protein, from the same livestock waste feedstocks. There are alternative routes from excreta to feeds under active consideration. For example, worms can be cultivated on cattle waste and harvested to produce a high protein meal as animal feed (Palz and Chartier, 1980); such applications involving worms or other invertebrate species as converters, comprise an important part of the 'bioplex concept' set out by Forster and Jones (1976) for recycling all wastes internally in an integrated food energy production system. It remains to be seen, once such procedures are developed, whether there is any energy value in the residues left from excreta after processing in this way.

E New approaches to utilisation of animal wastes: Solid/liquid separation

Slurry separators are already used on some farms to fractionate slurry into liquid and solid components. Working on cattle slurry, the separator yields, typically, about 20% by weight of a solid (with dry matter 16–22%) and 80% by weight of a liquid (with dry matter 4–8%). The option exists, therefore, to use one for fuel, and the other for feed, or to spread one on the land and utilise the other for feed, fuel or both. One of the most attractive options is to feed the whole slurry to the digesters for biogas production and then to fractionate the residues into a solid fuel component and a liquid fraction by means of the slurry separator, which, incidentally, operates at a greater throughput on digester residues than on raw slurry. The most proteinaceous particles are in fine suspension and separate with the liquid function, from which they may be removed by centrifugation. The solid fuel may be combusted on-site to heat the digester (thereby conserving biogas) and to dehydrate the solid fraction itself and the isolated protein feed component. The electricity required for operating the driers and centrifuges and for stirring the digester may or may not be generated from biogas. This overall scheme which seems likely to be capable of generating biogas at about £3.50/GJ in December 1979 terms, is illustrated in Fig. 3. The process, which is somewhat sophisticated, requires to be operated on a scale of 50 to 100 dry tonnes of feedstock per day, corresponding to gross gas production of 100 to 200 TJ/year.

Applications of such a process would result in the loss of whatever fraction of the feedstock nitrogen that would pass, with the solid fraction, to combustion. The work of Pain *et al.* (1978) on the operation of slurry separators suggests that this would be around 25% of the total feedstock nitrogen. If it were all valued at a price equivalent to fertiliser nitrogen, this would be worth £3.30/t of

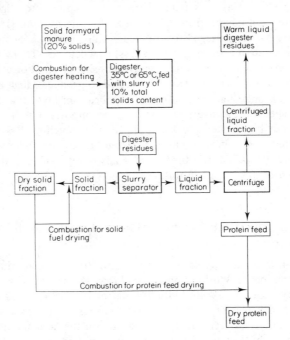

Fig. 3 Slurry separation scheme

feedstock solids at 1980 nitrogen prices, if the feedstock were slurry, but only about £1.60/t of feedstock solids if the feedstock were farmyard manure. In practice, much of this nitrogen is expected to be in an unavailable form and hence worth much less than this. It is worth noting, however, that the value of nitrogen, phosphate and potassium in cattle slurry, used as fertilisers, and only valuing the fraction that is available in the first year after application as a plant nutrient, amounts to approximately £14.30/dry tonne at 1980 prices; any processing scheme which fails to return most of this plant nutrient to the soil, will have to bear this amount as a cost; one may be able, however, to offset against this, the farmers' considerable costs for storage and spreading of slurries and manures.

[5]

Energy farming: Natural vegetation and dedicated energy crops

A The concept of energy farming

Chapters 3 and 4 have been confined to considering gathering the wastes from agriculture, as it is practiced today, and in those chapters the opportunities were discussed as if there were little or no possibility of adjusting the pattern of agriculture itself in accordance with the requirement to gather maximum biomass as fuel. Naturally, in introducing any such change one would expect to encounter great inertia; indeed, as soon as one is proposing to change the basic concepts, objectives and priorities in farming one may be thought of as disturbing the very roots of civilisation. Therefore, it is one of the most difficult areas involving biomass/energy schemes where the biomass requirement starts to make demands upon agriculture, beyond just collecting up its residues. The matter is emotive, because most farms have always been concerned primarily with food production and world food supplies are an emotive issue. It is emotive because, increasingly, farmers, the public and even, to a limited extent agricultural authorities, are starting to agree in some measure with the organic farming movement, that the organic carbon content of soils is an important parameter in determining their long-term fertility; the harvesting of fuel crops involves removal of carbon from the agricultural system for eventual combustion, instead of allowing it to re-enter the ground as humus. The matter could well become a political issue as soon as any major moves were made to implement energy farming, because people's lives would be affected, and so would the scenery. There could also be administrative problems to resolve in government, since in most countries, energy and food production come under different ministries, which would have to act in unison to secure any real progress with integrated systems for combined fuel and feed production. It is looking increasingly, however, as though these problems will have to be faced. So long as food production does not suffer, there seems to be little valid reason why agriculture should not produce energy also. The concern over the humus content of the soil is justified, but does not need to become a problem if the total output of organic matter from the land is sufficiently increased. The changes to scenery and lifestyles would not all be bad, and it looks as if governments may need to adjust the working of certain departments to cope with the situation.

The basic concept of energy farming stems from the recognition that ample sunlight energy falls on the Earth's surface each year to provide the food and energy requirements of the world population many hundred times over (about 3×10^{24} J reaches the Earth's surface each year, according to Slesser and Lewis, 1979) and that the living plant may be used to trap a small proportion of this, usually, for a good plant cover of the ground, between 0.1 and 5.0% of the energy reaching a given part of the surface in a year. In energy terms the main requirement is to maximise this capture of solar energy, and this is quite a different priority from the usual agricultural priorities. This priority may be modified by economic factors so that it becomes, for example, to maximise the economic production of readily harvested biomass. By contrast with this, the usual priority in agriculture is to maximise the economic production of particular food commodities, so the objective is at once restricted to a narrow range of edible plant species and, usually, to a particular part of the plant. A similar constraining of objectives may occur in biomass/energy if a particular type of fuel is needed, for example, ethanol. Until known and tested economic processes are available to produce ethanol or methanol from cellulose, alcohol plants will continue to require feedstocks rich in readily available carbohydrate, i.e. starch or sugars, as in the case of sugar cane, sugar beet, potatoes or cassava. In such a case, the biomass/energy scheme becomes more like agriculture in its growing and harvesting objectives; the high foreign exchange costs of liquid fuel for automobiles is already motivating some countries, notably Brazil, to think in these terms. In other cases the priority in biomass/energy is simply the highest possible yield of ash-free organic matter, almost regardless of its charactertistics, except that a premium value attaches to dry biomass, since almost the whole thermal content is then available as fuel, and that where biomass is wet, it is preferred in chemical forms that are amenable to microbial processing.

In most of NW Europe a massive biomass yield of about 2000 t/ha/yr dry matter would represent 100% utilisation of available solar energy. Since the most productive tropical crops peak at about 5% efficiency, a 100 t/ha/yr yield is the most that could be conceived in much of Europe, but the seasonal climate with low insolation and low temperatures in winter confine the maximum realistic objective to about 50 t/ha/yr, or 2.5% efficiency. In practice, yields of from 25–40 dry t/ha/yr have been observed with certain species of perennial weeds, by contrast with 6–8 dry t/ha/yr for typical arable crops in NW Europe. Although both sets of figures are variable, this four- to six-fold difference is at the basis of the principal case in favour of energy farming. On land that is not used for farming at present, one is free to try out these schemes. If sustainable, low-cost production of biomass at the levels indicated can be confirmed, it does seem likely that a pressure will develop upon agriculture to maximise the production of organic matter *per se* from even quite good quality farm land, and then to allocate appropriate fractions of the biomass yield to food, feed and fuel purposes. That is for the longer term, however. Before examining too closely the potential head-on collision between the food and energy priorities — and the possible resolution of that collision by a beneficial collaboration — it will be helpful to examine three simpler cases. The first is to consider almost irrespective of competition for land, to what extent annually grown agricultural crops could be dedicated to energy. The second is the role of catch crops, which are extra crops planted between main crops; the third is the role of perennial energy crops grown on non-agricultural land.

B Annual agricultural crops as dedicated energy crops

The obvious and commonly stated objection to using agricultural crops as fuel, is that spare land simply is not available to grow them without affecting food production. This is largely true in much of W. Europe, but it is not necessarily true in Australia, North and South America, and elsewhere. It is true that almost everywhere the best agricultural land has already been claimed but many places have reserves of ploughable, reasonably level land with productive potential.

In NW Europe the most likely crops to be used would be sugar beet, fodder, radish and forage crops such as clover, rye, lucerne or ryegrass, on account of their quite high annual productivity and moderate cultivation costs on a per tonne basis. The actual yields and costs experienced with each of these crops will vary with site and season, and the crops differ among themselves in performance and cultivation cost. However, a good result might well be a 70 t/yr harvestable fresh weight yield, 15 t/yr dry weight, at a 1980 cost of £460/ha/yr, £30.7/dry t or approximately £1.75/GJ of gross thermal value. After passing through a conversion process of, say, 60% efficiency, this leads to a feedstock cost (without processing), per useful GJ of converted fuel, of £2.87/GJ. At a minimum conversion cost at, probably, just over £1/GJ for wet biomass, a fuel cost of about £4/GJ results, comparable with oil costs at the time of writing. Costs above, then, do not preclude the use of arable crops for energy. In practice, the variation costs/GJ of gross thermal value can be expected; among the crops listed and, based upon costs for individual crops as cited by Spedding *et al.* (1979), this variation may well be over the range £1.70/GJ for red clover (in 1980 terms, allowing that Bathers' calculations were based upon Nix, 1977) up to £2.56 GJ for sugar beet. The latter figure gives a converted fuel cost of approximately £5.40/GJ, which is significantly more expensive, though even this, as a cost in 1980 money, is at a level that can be expected to compete as fuel prices rise over the next few years. However, in the case of sugar beet, its availability for energy is decided more by the price paid to the farmer for his crop than by his costs incurred in producing it. At 1979/80 prices for sugar beet roots in the UK an average yield is worth £765/ha/yr and the selling price for the whole crop, if the tops were included, represents £3.56/GJ of gross thermal value: using this price lifts the cost of converted fuel out of the reasonably competitive range to at least £7/GJ. Sugar beet, therefore, can be disregarded as an energy crop in NW Europe, or elsewhere with a similar population density, insolation and climate, except possibly as a source of premium fuels such as automobile fuel (ethanol). Potatoes command even higher prices.

The other potential arable energy crops listed show a more mixed composition, with lower ratios of sugar to fibre, making them less suited to fermentation to ethanol. Notwithstanding this, in New Zealand work is proceeding on the production of ethanol from fodder radish and also Jerusalem artichokes (Mulcock, 1978). On the other hand, the fermentable carbohydrate content of grasses, lucerne and clover are all too low to permit consideration of ethanol as a principal output; such crops would have to be considered for harvesting dry, if practicable, or otherwise as feedstocks to anaerobic digestion. In the case of fodder crops, the farmers' gross profit from their cultivation cannot be so readily assessed as with a cash crop such as sugar beet, because the crops are fed to animals and hence variables in the management of the livestock enterprise are essential to determine the profitability of crops grown. However, Spedding et al. (1979) calculates that forage crops of the type listed could be sold off the farm at prices which in 1980 terms, range from £1.75/GJ to £2.75/GJ of gross thermal value. These would represent the practicalbe selling prices for the crop as cut, and it would normally be in a fresh condition. After costs have been added for hay-making — the lowest cost drying method available — the material is not likely to be attractive as a bulky and inconvenient form of solid fuel. Chopping, baling, pelletising or briquetting of the hay are all possible, to give a more convenient form of fuel, but at the same time the costs incurred make it increasingly unattractive, either as solid fuel or as a feedstock to pyrolysis/gasification plants. On the other hand, as a feedstock to anaerobic digestion the crops do not need drying and gas which may be considered a premium fuel compared to solid fuel, could be available at from £4 to £6 per GJ. The conclusions may be drawn, therefore, that conventional annual agricultural crops do not appear favourable at all in NW Europe for solid fuels; root crops appear wholly unfavourable for any use except possibly, in the case of those with high readily available carbohydrate contents, conversion to ethanol (which will be examined more fully below), though their unfavourable economics are decided, not by the costs of growing them, but by the high price they command for their traditional uses; annually seeded forage species appear to be worth considering, though not outstandingly attractive, as feedstocks to anaerobic digestion.

Some of the forage crops being considered here can also be grown without annual cultivation and appear as potential species in section D on perennial plantation crops. In this role, their costs do not have to include the annual charge for preparing the ground and establishing the crop.

It can be seen from the foregoing that the economics of growing annually cultivated crops for energy can depend upon the entire economic environment in which the activity is to be conducted. Farmers would not divert sugar beet away from sugar manufacture towards energy in NW Europe because of the high price which the crop commands. Elsewhere, where there is no sugar industry or where there is plenty of spare land, the situation could be quite different, and energy production via sugar beet would then represent an opportunity for additional large scale farming enterprises to be set up, based upon taking low margins on very large volumes of extensively grown crop; if this were accompanied by higher levels of insolation than in NW Euope, with concomitant yield increases, this form of annual energy cropping could become really attractive. Such a situation clearly applies in Brazil, with excellent photosynthetic rates coupled with extensive land area related to its population, where sugar cane and cassava can be used for energy purposes, and to New Zealand, where annual crops of lucerne, sugar beet, maize, fodder beet, kale, rye and oats are being seriously considered for energy purposes (Stewart, 1978). From the Brazilian experience it is clear that annually cultivated crops can be viable for automotive fuel production given appropriate national circumstances and the New Zealand work and also Canadian work (Kirik, 1978) suggests that annual energy crops may well be viable for this and other purposes also. Much of more densely populated Europe cannot expect to derive the same benefits, lacking surplus land and, it seems, will be forced to investigate systems capable of increasing biomass output per hectare without incurring the annual cost of turning the land and establishing a new crop.

All projects involving annual agricultural crops incur greater expenditure of energy in cultivation than would be the case with perennial crops, on account of energy used in ploughing; very substantial applications of fertiliser are also required in most cases and manufacturing the nitrogen component of this calls for quite substantial energy inputs (e.g. see Table IV, chapter 3).

In respect of above-ground crops, these inputs are lowest for lucerne and clover because, as leguminous plants, they fix atmospheric nitrogen and the requirement for added nitrogen is minimal; in these cases input energy is about 5% of gross output. For ryegrass and cereals, which lack the advantage of nitrogen fixation, typical energy inputs are from 12 to 15% of output. Where the crop is a root crop, additional energy is expended on harvesting, and the range of inputs is from 15 to 20% of output. All these input figures are calculated on the basis of the gross thermal value of the crop; since in most cases a conversion process is needed, which is unlikely to be more than 60% efficient in converting gross thermal value into usable fuel, the energy inputs therefore range from about 8% of gross usable fuel produced, for lucerne, up to 33% for potatoes. Even these figures do not take account of the degree of fuel consumption in the conversion process; they would be appropriate where, for example, conversion was by anaerobic digestion, with 100% of gas being used for electricity generation and waste heat from the generator being used to heat the digester. However, where the gas itself was wanted for export from the site, the digester may require, say, 25% of gas to maintain operating temperature. In this situation the energy inputs for growing and harvesting become 11% of useful fuel output for lucerne and 45% for potatoes, with sugar beet, ryegrass and cereals falling between these two values. These figures really demonstrate the difficulty of using annual root crops as fuel, at least with the land availability situation and crop yields being as they are in NW Europe. Sugar beet appears to be the most nearly competitive root crop, but something approaching a doubling of yield, at least of average UK yield, appears to be necessary to make this crop attractive for general energy purposes by the time that energy inputs to its cultivation and the inefficiency of fuel conversion have been taken into account. On the other hand, grass and other forage crops may be worthy of further investigation for energy purposes, with the reservation that to minimise input energy, grass may need to be cultivated as a perennial rather than an annual crop and that, for the same reason, grass/clover swards seem likely to be more competitive than pure grass swards. This can be seen from results quoted by Gordon (1980) showing that a pure grass sward fed with 450 kg/ha of nitrogen, 90 kg/ha of P_2O_5 and 190 kg/ha of K_2O was consuming 38 GJ/ha of support energy, compared with a grass/clover sward receiving 60 kg/ha of nitrogen, 60 kg/ha of P_2O_5 and 140 kg/ha of K_2O, which was receiving 7 GJ/ha of support energy, assuming that the fertiliser is all put on in one application. Gordon's figures for energy recovered were quoted in terms of metabolisable energy, which is not the parameter required in questions of fuel production. However, working from his quoted yields of dry matter, it appears that the pure grass sward, which was giving the higher yield, required an extra 31 GJ/ha of support energy, compared with the grass/clover sward, to produce an extra 61 GJ/ha of gross thermal value in harvested crop. Clearly, this would not be worth doing unless the conversion of grass into usable fuel were more than 50% efficient. One may conclude from this that if grass were grown primarily for energy purposes, the supply of nitrogen from admixed clover is very well worth considering, even though the yield per hectare would be reduced.

Finally, in line with the conclusion already reached, that arable root crops in cool/temperate zones could only be considered for production of a premium fuel for which they were expecially well suited, such as ethanol as automotive fuel, it is of interest to quantify the costs and land requirements that are involved. Sugar beet is the most promising crop for this purpose, and as has been seen above, is available in NW Europe at a 1980 production cost of £2.56/GJ of gross thermal value, but a normal selling price from the farm of at least £3.56/GJ. To this basic cost must be added transportation, processing costs and the effect of conversion inefficiency, including an allowance for the fact that only the roots, rather than the whole plant would be useful for ethanol production. This leads to an acquisition cost of £7.12/GJ; adding £0.40/GJ for transport over a mean 15 miles to the factory gives delivered cost of £7.52/GJ; process efficiency in converting weight of sugar beet solids to weight of alcohol is approximately 63% but allowing for energy loss upon converting sugar to alcohol (7.3%) leads to adopting a reduced figure of 58%. Hence, the feedstock costs alone are of the order of £13/GJ of alcohol in 1980 terms, compared with a 1980 cost of petrol at the pumps of £10/GJ and process costs have yet to be added, together with transportation and distribution costs. Energy inputs to ethanol production are comparable with the energy outputs, so there is no doubt that the resulting automobile fuel would be very costly.

In view of the very large additions to costs for these items, alcohol from sugar beet seems most unpromising as a temperate biofuel, on cost grounds alone. The position seems even more unpromising when land requirements for growing the beet are considered in relation to European land resources, especially since good quality arable land is essential. At a high root dry matter yield of 7.5 t/ha, giving an alcohol yield of 4.3 t/ha (or 115 GJ/ha as liquid fuel), a project to produce in, for example, the European Communities, 5% of total energy requirements in this way (a figure of 2600×10^6 GJ/yr, which would make a substantial impact upon the provision of fuel for transportation) would call for about 22.6×10^6 ha or about half of all the Communities' arable land. Even then, no account has been taken of the requirements of energy for cultivation, harvesting and processing which, if they had to be obtained from beet itself, would come close to demanding 100% of the available arable land. This really illustrates the hopelessness of thinking along these lines for provision of anything more than a tiny proportion of energy requirements in a land-hungry, densely populated, industrialised temperate area. The solutions that find a ready application in Brazil apparently have little place in Europe. Brazil's relative success with alcohol production from sugar cane and cassava is related, first and foremost, to the high photosynthetic rates achievable in tropical climates, about 90 t/ha/yr for sugar cane, (Jackson, 1976) and the readily availability of land for plantations of cassava in areas which, although the standard of the soils may be mediocre to poor, are not competed for by other crops. In this situation farmers' costs per tonne are lower and margins per hectare can be reduced and still be acceptable to the farmer engaged in cultivating a very extensive land area for bulk deliveries of a single crop.

C Catch crops

A catch crop is generally sown during the summer months, following the harvest of a main crop, for harvesting before the next main crop is due to be sown. Therefore, it is a way of utilising agricultural land during a period when it would otherwise be vacant and unproductive. Some farmers already grow catch crops, without any fuel connotations, and usually apply them as animal feed. A few organic farmers grow catch crops to plough in, as a source of 'green manure'. It appears to be true, however, that most catch crop opportunities are missed for one reason or another; quite often, arable farmers simply do not have livestock to feed. This opens the possibility that farmers who could grow catch crops but do not choose to do so at present could be induced to grow them if an organised system of contracting existed to furnish them to energy plants as feedstock.

A study on catch crop opportunities in the UK has been undertaken by Spedding *et al.* (1980). One of the most crucial variables is the length of growing period which is available for the catch crop; when this period is very short, the crop is scarcely established before it must be harvested to prepare the ground for the next main crop. Yields are then very low and all the expenses of ground preparation, sowing and harvesting are committed in relation to a very low return. Where the main crop is early vegetables the land may become available in mid-June, for example; significant growth is possible up to about 1 November, giving a maximum growing period of about 19 weeks. However, this opportunity applies to such a small proportion of the arable land as to be quite insignificant nationally, although the yield of catch crop dry matter per hectare would be good. Harvesting of major agricultural crops commences about late July and, with some crops such as sugar beet, continues well into the winter months. Land which becomes free after 1 October is not considered from the point of view of catch crops because there would be less than four weeks growth before the virtual end of the growing season. The land becoming available between late July and 1 October comprises land which has been main-cropped with cereals, rape, peas and beans, potatoes and maize; additionally, the very first harvesting of sugar beet may just be early enough to allow a minimum period for catch-crop growth. Quantitatively, by far the most important opportunity is created after the cereal harvest, which is generally concentrated within the month of August. An analysis based upon the data of Spedding *et al.* (1980) was carried out by Plaskett (1980) and by Carruthers (1980) and indicated that, at least in the UK the mean period of land availability was quite short.

The available land can be usefully divided into one category that is required again for replanting with main crop in the same season, usually winter cereals, and another which is not scheduled for replanting until the following spring. According to Carruthers (1980), the mean period of availability for the first of these categories, in the UK, is 5.47 weeks and for the second is 8.33 weeks. However, 77.6% of the land falls into the first category, giving an overall mean of 6.11 weeks; allowance of about 1 week must also be made for clearing up after the preceding main crop, preparing the ground and fertilising and, at the end, for preparing the ground in readiness again for the following main crop. Therefore, the average period of availability for catch crop growth in the UK is close to 5 weeks, and there is no doubt that the economics of obtaining energy by this method in the UK are constrained by the rather limited growth that could be expected during this period. Spedding *et al.* (1980) considered three of the ITE* Land Classes (Classes 3, 11 and 12) and concluded that these might be expected, on average, to yield 71.5, 61.2 and 77.2 GJ/yr respectively as gross thermal value of catch crop material, implying dry matter yields of 3.5t/yr up to 4.4t/yr and leading to mean production costs in the range £2.63 to £3.31/gross GJ of unconverted fuel. Taking these figures leads to a feedstock cost per GJ of converted fuel of £4.38 to £5.52 without considering process cost, feedstock transportation cost or farmers' margin. In practice, it seems doubtful whether such yields could be obtained in a 5 weeks period; the data of Austin (1974) suggests that a dry matter increment of 0.5 t/ha/week is the most that can be expected from catch crops during the early period of growth, although the weekly increment can exceed this considerably in the case of catch crops which have been already growing for ten or more weeks, when a rather complete leaf canopy has been formed. As soon as the expected dry matter yield of only 2.5 t/ha/yr is considered for a 5 week growing period, the harvest is only 44 GJ gross energy and costs per GJ rise to £4.61/GJ of gross thermal value and feed stock would be costing £7.69 per GJ of converted fuel, a price which already appears uncompetitive, without adding process costs, transportation and farmers' margin. The picture would be different in different situations, of course, so that the 22.4% of the land that has a mean period of availability of 8.33 weeks (see above) might well grow catch crops for 7.33 weeks, provide a yield of 3.66 t/yr, costing £3.17 per GJ of gross thermal value, or £5.28 per GJ of converted fuel, for feedstock alone. Even this seems too expensive to be practicable, so in the UK it is likely that energy catch crops will be relegated to that very small proportion of the arable land that is available for substantially longer than 8 weeks, at least until the cost of fossil fuels has doubled or trebled in real terms compared to 1980 prices, or until improved techniques are introduced to improve the yields that can be obtained.

The two principal opportunities for increasing catch crop yields are through whole crop harvesting and aerial sowing.

Whole crop harvesting has been mentioned already in chapter 3 and comprises in the case of a cereal crop, cutting the whole crop without threshing and carting to a factory, where it is dried and separated into a variety of useful fractions. Introduction of such a method would influence catch crop potential by enabling the preceding cereals to be harvested a little earlier, typically about three weeks earlier; the reason for this is that a drying operation is usually included in the whole crop harvesting concept and it matters very little if the grain is harvested in a slightly wetter condition than is usual by conventional procedures. From the point of view of catch crop production, an extra three weeks growth might be gained for the catch crop, worth an addition of about 1.5 t/yr of dry matter yield.

*Institute of Terrestrial Ecology

Aerial sowing is significant because it would enable the catch crop seed to be sown while the cereal crop is still standing in the field. this can be done about three weeks before harvest; when harvest takes place the catch crops have already germinated but are too small to be cut with the cereal harvester, and they are left to grow on immediately after the cereal has been removed. A combination of aerial sowing and whole crop harvesting of cereals is quite likely to add approximately six weeks to the growing period available for catch crops. This would result in a mean period of $10\frac{1}{2}$ weeks on that part of the available land that is required again in the same season, giving a probable yield of 5.7 t/ha/yr and 13.33 weeks on that land that is not required again in the same season, giving a probable yield of 7.2 t/ha/yr. The production costs of these would clearly be lower than those calculated above, for catch crops conventionally sown following conventional harvest. The two yield figures given above of 5.7 and 7.2 t/ha represent 100 and 126 GJ/ha/yr respectively, and lead to probable costs of £2.00 and £1.60 respectively, per GJ of gross thermal value, if the cost of establishment and harvesting are unchanged, or £1.60 and £1.29 respectively if, as suggested by Spedding et al., these costs are reduced by 20% on account of aerial sowing. The latter figures lead to a feedstock cost of £2.67 and £2.15 respectively per GJ of converted fuel, at a conversion efficiency of 60%. These must represent the most optimistic projections that can reasonably be made in the UK at the present time for energy catch crops, and indicate that catch crops can perhaps prove competitive as a source of fuel if whole crop harvesting and aerial sowing are used, though without these innovations their future appears rather bleak on economic grounds, since costs per GJ are directly linked to the absolute yield of dry matter that can be found during the growing period.

Species of catch crops that appear promising in temperate agriculture include fodder, radish, pea, vetch, rape and beans; kale and fodder beet are associated with higher establishment and/or harvesting costs.

Palz and Chartier (1980) calculated the maximum potential energy contribution from catch crops in the UK and, by extrapolation from data on arable area and climate, in other countries of Europe. The UK estimate was 320×10^6 GJ, or 7.27 Mt oil equivalent. However, this estimate was based upon most of the tillage area being available and upon a 50% introduction of aerial sowing and whole crop harvesting, and takes no specific account of environmental and cultivational constraints. Even so, it is not very different from a detailed assessment carried out by Spedding et al. (1980) which arrived at 214×10^6 GJ for the UK, by conventional cereal harvesting and post-harvest sowing, or 311×10^6 GJ if aerial sowing (but not whole crop harvesting) was carried out everywhere. These figures equate with 4.86 and 7.07 Mt oil equivalent, or 2.4 and 3.5% respectively of national primary energy consumption; fuel conversion efficiency at 60% reduces these figures to 2.92 and 4.24 Mt oil equivalent respectively and to 1.4 and 2.1% respectively in national primary energy consumption. The extrapolations in Palz and Chartier (1980) to give catch crop contributions in other European countries, give France, 25.25; W. Germany 12.30 and Italy 12.18 Mt oil equivalent, although in Italy there is considerable room to doubt the ability of farmers to establish a catch crop during the dry summer period, except on irrigated land. These French, W. German and Italian figures require to be multiplied by 0.6 to give the amounts of usable fuel coming from a typical conversion process.

D Perennial non-woody plantation crops

The main reasons for considering perennial plantation crops are (*i*) that if the crop does not have to be re-established each year, cultivation costs should be much lower than for annual crops, (*ii*) that a perennial crop, with its root systems well established in the soil and nutrients stored underground in roots and tubers, is able to send up shoots as soon as the weather is warm enough to form a canopy for collection of solar radiation efficiently, even during the first weeks of spring. The early establishment of good ground cover can be expected to lead to an increase in photosynthetic efficiency, measured over the year, compared to annual plants that have to establish their root systems first, (*iii*) as perennial plants are often tall and woody, there is reason to expect that stem material, which is relatively nutrient-poor, will comprise an important part of the biomass formed, leading to a higher biomass production per unit of nutrients applied than in the case of annual herbaceous species, (*iv*) some wild perennial species are believed to survive and grow well on land having very low levels of available nutrients which would be hopelessly inadequate for arable farming; for example, this appears to be the case with bracken (*Pteridium aquilinum*), which is well able to utilise soils with very low potassium content (Aston, 1917). This gives rise to the expectation that such plants will perform well as energy crops in poor soils and reinforces the impression gained by considering their leaf/stem ratios, (*v*) higher ratios of output to input energy than is the case for agricultural crops; clearly, this parameter is linked to limited energy expenditure on cultivation and fertilisers.

These considerations have resulted in evaluation of wild plants, exotic foreign perennials and perennial weeds as potential energy crops which, once established over an area of ground would produce large biomass yields year after year, excluding other invading species by the density and vigour of their own growth, with the help of minimal management action. For example, bracken, as mentioned above, is highly productive, and up to 14.1 dry tonnes/ha has been measured in the British Isles (Pearsall and Horham, 1956). While this is not greater than the best yields obtainable from productive agricultural species such as ryegrass and sugar beet, the potential advantage of bracken as an energy crop would lie in potentially very low production costs, in both money and energy, and the use of land quite unsuited to any normal form of agriculture apart from low-grade grazing. An isolated report (Aston, 1917) gives 60 dry tonnes of bracken as an annual production rate at a site in New Zealand, although this measurement has not been repeated and confirmed. Gorse (*Ulex europaeus*) was also found to be very productive in New Zealand (Egunjobi, 1971) giving almost 17 t/ha/yr of dry material and a reed, *Phragmites communis*, which grows in lakes, was found to give almost 11 t/ha/yr by Mason and Bryant (1975). *Cynodon dactylon* was found by Burton *et al.* (1959) to yield 26.3 t/ha/yr in Georgia, USA. A range of other species have been studied by Callaghan *et al.* (1978), especially *Polygonum*, *Spartina*, *Impatiens*, *Epilobium*, *Gunnera* and *Heracleum*, as well as other species. Stinging nettles, *Urtica*, have been found capable of yielding up to 10 dry tonnes/yr, but do not respond well to regular cutting, suggesting that either the plant is incapable, physiologically, of withstanding such treatment or that the nutrient loss from the foliage-rich crop is too severe. Interest centres particularly upon species which may produce more dry matter per hectare than agricultural crops, particularly *Polygonum* species that have been shown, in natural stands, to yield 25 t/ha/yr, or, in one case, almost 40 t/ha/yr dry weight. If such yields could be sustained from year to year, these plants would assume an immense potential importance in future energy scenarios. At present, this is not known and it is also uncertain how much fertiliser would be required, over and above the possible spreading of digester residues, to enable the production rate to be sustained. As in the case of stinging nettles, one has to distinguish the plant's inherent physiological ability to respond by rigorous regrowth after being repeatedly cut, regardless of the nutrients present, from the nutrient-dependent aspects of regrowth. Investigations are needed on these, and other promising plant species, to assess their inherent potential for vigorous regrowth at the same productive levels, given nutrients in more than adequate supply, because any physiological barrier to regrowth may be difficult to correct, making a species unsuitable. Once vigorously regrowing species are established, the requirements of a plantation for nutrient recycle and nutrient supplementation becomes of major concern. Another, possibly important, aspect is the timing of cutting and the number of cuts per year. Perennial crops in general may be cut (*i*) several times a year, (*ii*) once, in late summer, when the dry-matter present in harvestable portions of the plant are at a maximum or (*iii*) once in autumn, when the sap and much of the nutrients in the above-ground parts have been withdrawn into the roots or rhizomes for storage until the next spring. The latter certainly happens with bracken and *Polygonum*, leaving fairly dry, fibrous brown stems available for harvest at that time. The ability to withstand several cuts per year depends, if nutrients are adequate, upon the inherent physiology of the plant; fairly frequent cutting is a good way to secure maximum yields from grasses, but cutting during the spring or early summer is recognised as a way of eradicating bracken. Among tropical crops, it is possible to cut cassava repeatedly to obtain good yields of the tops, which are then obtained in a green and tender condition compared with the woody tops of matured cassava, but if this is done good yields of roots are naturally not obtainable at the same time. Although stinging nettles are notorious for their vigorous regrowth when they are cut in an effort to eradicate them, they appear unable to sustain the same productivity when cut regularly. The ability of *Polygonum cuspidatum* to withstand frequent cutting, or even just annual cutting, with maintenance of yield, is not known at the present time, and virtually nothing is known of the best means for harvesting a really enormous size, prolific species such as *Gunnera manicata* (Lawson et al., 1980). Harvesting *Polygonum* or bracken in autumn gives a product with relatively low moisture content, which could presumably be burnt directly or processed into solid fuels in the same way as straw, and bracken is already harvested in this condition by some hill farmers for use as livestock bedding. The yield will clearly be less under these conditions than by harvesting the same plants when they are actively growing, or at the point where harvestable biomass reaches a peak, due to the withdrawal of sap and nutrients into the underground portions of the plant; in such material, fibre can be expected to form a higher proportion of total solids, while metabolisable energy and organic solids convertible by bacteria into methane, will be minimal. Much work is needed to find out whether these plants can withstand the removal of their above ground parts, even once a year, together with all the nutrients they contain, either physiologically or nutritionally. A case can be made that off-take of nutrients should be of little consequence provided these nutrients are replaced, as they can be, in the form of digester residues, leading to a complete recyclying system; higher off-take of nutrients in the harvested biomass should mean, in these circumstances, only a faster cycling of nutrients without necessarily inducing any losses from the system. In practice, however, nutrients already present in the plant body (i.e. the underground parts) are more immediately available and more free from loss-causing effects, such as leaching, volatilisation etc. compared to nutrients in digester residues applied to the surface of the soil; even if losses do not occur and are in a chemical form which is available for reabsorption, there will be a time lag before 100% of them actually come into physical contact with the root hairs and are absorbed. If experimental work shows that harvesting at peak biomass is not compatible with high yields sustainable from year to year, it may be necessary to accept only the autumn harvest of dry fibrous matter and exploit the advantages of its relative dryness and fitness for direct combustion.

No work has yet been done on the establishment costs of plantations of perennial energy crops and no-one knows how much management attention will be required during succeeding years to keep the plantation as a monoculture and maintain its productivity. The chances are that levels of cost for initial establishment will be very species specific; it is certain that a species such as comfrey (*Symphytum asperrimum*) which requires to be propogated by individual cuttings, will always be very expensive, and figures quoted by Hills (1976) extrapolate to a 1980 establishment cost of about £1250/ha. If this cost is spread over, say a 10-year life of the stand, with annual management costs still to be added, the total annual costs would probably approximate to those of an annual farm crop.

This seems too expensive for fuel purposes. However, perennials such as *Polygonum* can be established by burying lengths of the rhizome and, although these would have to be produced by digging up stands of *Polygonum* and chopping the rhizomes into suitable size lengths, this procedure should not be more costly than the production of seed potatoes. Since one can also imagine *Polygonum* rhizomes being planted by machines similar to potato-planters, the overall costs for 'seed' and labour in establishing this crop may be broadly similar to potato, perhaps in the region of £650/ha. At this point, assuming that the *Polygonum* stand will last longer than 5 years before replanting, and taking into account the very high yields hoped for from this plant, the prospects of such plantations for energy purposes begin to look attractive. The life of stands might eventually be extended by procedures developed to resuscitate them or by simply ploughing the fields. Obviously, any such perennial crop that could be established by seed would have an enormous advantage, since its establishment costs would then be similar to those of most arable annual crops, but capable of being spread over many years. By assuming some fairly standard cost levels for operations such as harvesting, pest treatments, rent and rates and the spreading of digester residues, together with a fertiliser requirement similar to that of sugar beet, production costs of between £0.75 and £1.32/GJ gross thermal value; in 1980 terms, can be forecast for *Polygonum* yielding 25 dry t/ha/yr, or £1.07/GJ gross thermal value if only 17.5 GJ/ha/yr are obtained, all dependent upon the assumed cost relatively between *Polygonum* and potato for establishment. This range leads to a feedstock cost per GJ of gross gas production of between £1.70 and £3.00 and a minimum cost for converted fuel after transportation and conversion costs of between £2.80 and £4.20/GJ in 1980 terms, although farmers' margins have not yet been considered. This range of costs must be considered potentially competitive in relation to the likely fuel cost structures of the fairly near future, even without the possible added benefit of deriving animal feeds from the digester residues.

It is of interest to assess the potential contribution from perennial non-woody plantation crops to national energy economics. In any country with major reserves of ploughable, though low grade agricultural land, the opportunity is likely to be a major one, the most obvious competitor for bio-energy generation being forestry, and the choice would be made on the basis of relative productivities and upon the particular chemical and physical form in which the fuel was

ENERGY FARMING

required. In densely populated countries, this opportunity, as, indeed, any biofuel opportunity, will be limited in scope by land availability and is not likely to approach total supply of the present energy requirements, contributions being quite meagre so long as agricultural land is not utilised. For example, in the UK the allocation of one million hectares to perennial energy crops on non-agricultural or only the poorest rough grazing land is, perhaps, the most that is credible, bearing in mind the demands of recreation for land and the difficult cultivability of much of the land in question. If a dry matter yield of 25 t/ha/yr were achieved over this whole area, a high demand in view of the character of the land, the harvested biomass would total 25 Mt/yr, 440 M GJ and converted fuel at 60% efficiency would amount to 264 M GJ, 6 Mt of oil equivalent, or about 3% of national energy requirement. Similar constraints would be met, in greater or lesser degree, in most countries of Europe. However, an improved prospect develops if agricultural land is brought into consideration. At least 13.8 M ha of the UK, or about 57% of the land area, is grassland, mostly being applied to some degree of agricultural use. Average British grassland yields approximately 6.5 dry tonnes/ha/yr, according to Cooper and Breese (1971). Rough grazing is not usually considered as 'grassland' agriculturally; exclusion of rough grazing brings the total area down to 7.3 M ha and an average yield on this of 6.5 dry t/ha/yr. produces 47.5 M t/yr of grass dry matter, equal to 830 M GJ/yr, which can be presumed fed to stock under present conditions. If this much forage (or slightly more to compensate for the contribution from rough grazing, say, 1200 M GJ/yr) could be produced in the form of the leafy fraction of an energy crop grown on the land which is now grassland, there would presumably be no obstacle, in the extreme case, to taking over the whole grassland area, including rough grazing, for perennial energy crop production. A 25 t/ha/yr yield on 13.8 M ha amounts to some 6000 M GJ, and provision of the forage requirement, so as not to interfere with production of livestock and livestock products, would leave 4800 M GJ for energy purposes, 2900 M GJ after conversion, 65 M t of oil equivalent, or 30% of national primary energy demand. The British scenery would change, as would the employment situation within agriculture, but a really major contribution to energy would have come from biomass, notwithstanding the small area/high population density situation of the country. It is conceivable that the UK, which has on the whole quite a spendthrift attitude to the consumption of fossil fuels, could manage to maintain living standards on half the present per capita energy consumption and if that were done, the country would be another step closer to permanent energy independence. Biomass could still be forthcoming from crop wastes, livestock wastes, non-agricultural land, forests etc, so a scenario in which biomass becomes the major energy source is not entirely inconceivable, even in the UK. European countries with larger ratio of area to population, as in the case of France, would be in correspondingly stronger position. No-one is suggesting that the means is available now to make this scenario come to reality, but the chances of success appear high enough to stake a great deal upon the necessary crop breeding programmes, crop fractionation studies and energy conversion technologies.

Finally, coverage of this topic would be quite incomplete without mention of those tree species that are capable of growing in dry arid areas to produce non-woody biomass. Since some of them are trees they are, in one sense, encompassed within forestry (see chapters 6 and 7); indeed the tree genera involved, such as *Leucaena* and *Prosopis* can and do yield wood for fuel in quantities on much land that would generally be considered most unpromising for growing biomass of any kind (National Academy of Sciences, 1977; Felker, 1979). However, *Leucaena*, intensely cropped and never permitted to grow to tree size, may yield 2–20 t/ha dry matter in the form of leafy material suited to anaerobic digestion and/or cattle feed, while *Prosopis* may yield sugar-rich pods of the order of 4–10 t/ha/yr on land which is otherwise useless for agricultural purposes, such as the Atacama Desert of Chile, where rain often does not fall for several years at a time. *Prosopis* pods are often utilised directly as food for ruminants; however, it seems likely that they could equally well be utilised for energy in the form of biogas or ethanol, with the proteinaceous residues being fed to pigs or poultry. *Prosopis* is the more drought-resistant genus of the two, but *Leucaena* can flourish without rain for up to two thirds of the year and with total annual rainfall of only 25 cm. *Euphorbia* is a drought-resistant succulent genus which, like the rubber tree, *Hevea brasiliensis* produces isoprenoid hydrocarbons (see Table III, chapter 2); although these are based upon the 5-carbon, partially unsaturated branched isoprene skeleton and are polymerised to a molecular weight of 10 000–50 000, the suggestion has been made that these plants, especially *E. lathyrus* and *E. tirucalli*, could be cultivated on arid land, the yields of hydrocarbons being maximised by genetic manipulation, and the products depolymerised to yield a substitute petrol.

This is therefore, the most exciting long-term prospect of developing the cultivation and genetics of these plant types to make large woody and non-woody biomass contribution in extensive areas of the world having high insolation rates but low water supply.

Because of the great potential importance of this subject, attention is now given below to the technological approach to fractionating energy crops (whether they be annual, perennial, catch crops or natural vegetation) into feed and fuel fractions.

E Processing options for non-woody biomass: Integrated feed/fuel production systems

While the concept of separating non-woody biomass into fractions for the different uses has special relevance to perennial plantation crops on account of the way in which it may make agricultural land available for them, the same concept may well be applicable to catch crop material, annual crops, crop residues and even the harvest from natural vegetation, if fractionation results in more profitable utilisation.

The simplest form of fractionation is physical separation and sorting of different parts of the plant. This can take the form of either macro- or micro-separation. Sorting of Brussel sprout waste into stalk and leaves, for different uses, would be an example of macro-separation, as would the separation of the leaves, shoots and tender upper stems from the thick fibrous stems of a perennial plantation crop like *Polygonum*. In the latter case, the leaves and tender stems may well be appropriate for fodder to ruminant animals, and hence be of a higher value than the fibrous stems, which would be used as fuel. A preliminary assessment of *Polygonum* (Plaskett, 1979, unpublished) shows these two fractions to be apportioned in the ratio of 1:2, leading to a fodder yield of over 8 dry t/ha/yr from a stand that achieved 25 t/ha/yr total dry weight. This fodder fraction would support ruminants rather than non-ruminants; this would be in accordance with its conceivable role as a replacement for grass (although it is clearly possible that a combination of several perennial species, or even specially bred varieties may be needed to achieve this goal fully), and no contribution would be made to the pig or poultry industries. Such a contribution becomes conceivable if micro-separation is introduced, as has already been applied to lucerne, whereby a mechanical process strips the leaves of the 'flesh' between the veins, and the fibrous leaf skeleton remains; such treatment obviously results in a much reduced quantity of feed at a considerably higher protein content since, within the already fairly proteinaceous leafy fraction, it is the cells of the 'flesh' or parenchyma between the veins which contain most of the protein, these being the most actively metabolising cells of the leaf. Such material may reach about 30% or more in crude protein and, being very low in fibre, may be suited to the needs of non-ruminant species. Most European countries import considerable quantities of protein concentrates in the form of soya meal, fish meal or meat meal, for feeding to both ruminant and non-ruminant livestock; the possible future availability of huge plantations of perennial energy crops might, therefore, open the way to this being replaced from indigenous sources, without necessarily having to resort to the greater complexity and energy expenditure of juice expression from the crop. Even with macro-mechanical separation, major energy expenditure questions arise when it comes to preserving the crop from the time of harvest so that it may be made available to the animals throughout the year. Drying the bulky ruminant fodder seems to be ruled out by the very high energy requirement of the process, amounting to 8–13 GJ/t, depending upon the moisture content of the dryer feed, being roughly equal to the energy available in the fodder itself after allowing for losses during the conversion to usable fuel. Drying the much smaller amount of micro-separated non-ruminant protein feed may be justifiable, especially if cheap solid fuel is available at the energy plant, derived from the perennial crop itself, in view of the smaller drying load involved and the higher market value of the dry product. For the bulky fodder fraction, destined for ruminant animals, ensilement appears the best solution to preservation; by concentrating the housing of these animals around the energy plant itself, transportation of this material would be minimised, and the animal wastes would also arise where they could be immediately utilised for fuel production.

A more thorough fractionation would require juice expression in accordance with the proposals often put forward by Pirie since the 1940s. (Pirie, 1966, 1971) and since then widely advocated and discussed for agricultural purposes, involving ruminant and non-ruminant feeding (Green Europe, 1974; Wilkins *et al.*, 1977) and as an aid to crop-drying (Davys, 1974). Wilkins *et al.* (1977) concluded that, agriculturally, the process did not offer adequate returns to provide an incentive for the necessary investments, and Spedding (1977) considered this to be largely attributable to the high energy requirements of the process. However, no assessment has yet been made of this form of crop fractionation in relation to combined agricultural/energy utilisation; the massive production of the biomass that provides the feedstock, together with avoidance of dehydration steps wherever possible, may alter the economic prospect. Based upon the energy cost figures presented by Wilkins *et al.* (1977) for fractionation of specially grown lucerne, it appears likely that only about 8% of the gross thermal value of the crop was required for fractionation into juice and wet pressed fodder and for coagulating, separating and drying the protein, although this would escalate to 13–14% in terms of converted fuel. When, as is being suggested here for energy crops, only a third of the crop is fractionated in this way, the energy cost seems likely to drop to less than 5% of the gross biogas potentially obtainable by digesting the whole crop, even after allowing for energy expenditure in separating the fuel and fodder fractions, and should be acceptable in order to gain optimum utilisation of the harvested material.

The principle of crop fractionation by juice expression is that the juice, which can be obtained in copious amounts from moist green vegetation, contains dissolved and suspended protein and heating to 90 °C coagulates this protein, so that it becomes concentrated in a small amount of separable coagulum, having a protein content of between 50% and 80%, depending upon the composition of the plants being used and depending to what extent the coagulum is washed. Since virtually all the fibre is left behind in the pressed residue, the dried coagulum makes an excellent protein concentrate for feeding to pigs and poultry and, in the case of poultry an added bonus is obtained in the form of the carotene in the coagulum, which serves to colour egg yolks and the flesh of broilers. In its strictly agricultural version, protein is not exhaustively extracted from the crop because enough must be left to ensure adequate nutrition of the ruminant animals that will eat it. The process therefore exploits the surplus protein present in forage over and above nutrient requirements. In practice this means that a large volume of forage has to be lightly processed to 'cream off' this few percent of surplus protein, effecting a reduction, for example, from 18% to 14% crude protein, and this necessarily has its impact upon the relativity between processing costs and the value of the protein gained. As far as the writers know, no attempt has been made yet to develop a more economic version of this process by extracting protein from 50% of the crop exhaustively, and mixing the protein-depleted residue 50:50 with fresh unprocessed crop, though at first sight this would appear to be quite a promising approach. However, in a process version designed to produce non-ruminant protein feed and energy, no limitations of this sort would apply and the protein step could be conducted as exhaustively as was compatible with process economics, leaving protein-depleted, high fibre residue to be used as fuel. If the non-ruminant animals that were to consume the protein were housed around the energy plant, even the need to expend energy on drying the protein might be avoided by employing the method investigated by Subba Rao *et al.* (1967) for preserving wet leaf protein concentrates with acetic acid (organic acids in silage run-off would be worth testing for the same effect) or, to adopt a method which does not even incur the energy cost of protein coagulation, the method investigated by Stahmann (1977) which entails keeping uncoagulated juice anaerobic until acidification occurs through microbial action, and the protein precipitates; it is not clear what factors prevent attack on the protein by juice protease enzymes during the fermentation period, but reliable coagulation of the protein was reported. There are, therefore, several combinations of ways in which fractionation by juice expression could be applied to energy crops, while maximising protein and energy yields and minimising energy consumption in processing, making it attractive for the future to investigate this process in some depth.

Fractionation by juice expression could be applied to:
the whole crop, without prior mechanical separation;
the forage fraction only;
the stem fraction only;
both the forage and stem fractions separately;
conceivably to the micro-fractionated leaf parenchyma fraction only, with or without separate juice expression from the rest of the crop.

All these different actions would be expected to have different outcomes, and the possible advantage of juice expression from the stem fraction would be to take advantage of its distinct composition. Such a juice may well be protein-poor but high in sugars; for example, if 10% of the crop dry matter could be obtained in this form, rich in sugar that was convertible to ethanol, this would appear to be a more promising approach to producing ethanol for motor fuel than the growing of special annual arable farm crops for the purpose. The brown juice

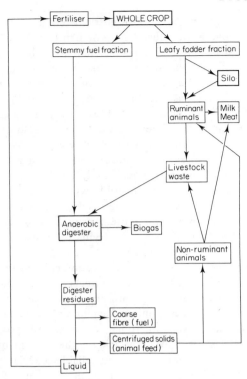

Fig. 1 Simple energy crop fractionation scheme, not involving juice expression or micro-separation

left after coagulating protein from the leafy fractions is also sugary and could be treated in the same way. If this were not economic, methane could certainly be generated very rapidly from the diffusable solutes of these juices, leading to low detention times in the digester and low digester capital costs for a given output, leaving the fibrous fraction to be used as solid fuel. Various versions of these integrated process options are given in Figs 1 and 2.

F Harvesting natural vegetation

An alternative to establishing plantations and managing them is to simply harvest the natural flora. One must expect that with this option, the yields will be far lower, but so also are the input costs, being confined to harvesting and transporting the material. It appears to be a very real option wherever the land is not ploughable, or where particularly productive stands of suitable species have already established themselves naturally. In most countries there is a distinct

ENERGY FARMING

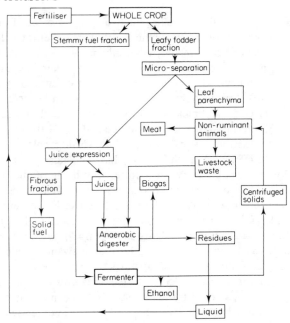

Fig. 2 Energy crop fractionation scheme involving micro-separation and juice expression

lack of information about the extent and character of natural vegetation, and this stands in the way of estimating its productive capacity. In the UK, the area of natural and semi-natural vegetation, if it is derived by subtracting the cultivated area from total rural area, still includes 'rough grazing', which may vary in quality and type, from previously cultivated grassland which has been neglected, to, at the other extreme, bare rock. There is even quite a considerable doubt about how much such land there is, the conflicting nature of various estimates have been reviewed by Lawson et al. (1980), and these authors have made estimates of the percentage of rural Britain covered by the most widespread indigenous species or species groups, and concluded that when this estimate is completed by taking account of all indigenous species, it would exceed the official estimate of 'natural and semi-natural vegetation' of 37.8%. They found heather (*Calluna vulgaris*) to be the most widespread, covering 7.12% of the rural area; purple moor grass (*Molinia caerulea*) covered an estimated 3.33%, bracken (*Pteridium aquilinum*), 1.54% and common bent grass (*Agrostis tenuis*), 3.81%. At an estimated 20-year harvest yield of 16 t/ha dry matter on the 1.6 M ha of land in Britain covered by heather, the total annual yield for the country could be in the region of 1.28 M t/yr, or about 0.5 M t of oil equivalent, though in practice, some steep or rocky areas would doubtless prove unharvestable. The 0.34 M ha covered by bracken, at 11 t/ha/yr, might yield, at most, 1.5 Mt of oil equivalent/yr. These are worthwhile quantities, which could be taken at a low moisture content, and should not be neglected, especially since they originate from land which it might be difficult to render productive in any other way, unless forestry was a practical possibility, both from the standpoint of agronomy and politics. Lawson et al. (1980) consider that heather biomass might be available for a cost of less than £0–05/GJ, and bracken biomass from natural stands at less than £0–20/GJ of gross thermal value. This would be just about the cheapest biomass fuel, cheaper even than municipal waste when sorting costs are considered. Callaghan et al. (1979) have pointed out that they also have a particularly favourable ratio of output energy, to input energy being of the order of 100–200, compared to values of 6–16 for grass and 3–12 for field crops.

Natural vegetation in wet places should not be overlooked entirely, though the areas available are usually modest. Sphagnum moss can yield up to 2 t/ha/yr (Forrest and Smith, 1975), marsh plants such as *Spartina* and *Limomium* may produce around 10 t/ha/yr (Jefferies, 1972) and swamp reeds can be very productive, being continually bathed in nutrient-rich water, with for example, *Phragmites* or *Typha* reaching 10–15 t/ha/yr. *Phragmites* is of particular interest because a project has been recently undertaken in Sweden to harvest this species from lakes and to apply it to fuel use. The work has been extended to artificially planting the reed rhizome in parts of the lake where it did not grow naturally; this then represents a combination of natural vegetation harvesting with the energy plantation concept. It seems likely that in practice these two concepts will often merge to some extent, with varying degrees of management being applied within the context of the natural habitat of the species concerned.

G Summary of productivities and costs

A summary of productivities and costs for plant material for bio-energy purposes is given in Table I, to enable easy visual comparison of figures presented in chapter 3 and in other sections of this chapter.

Table I Vegetable biomass: Summary of productivities and costs (1980)

Type of biomass	Most likely range of yield (dry t/ha/yr)	Probable range of costs (£/GJ gross thermal value, before transportation)
Vegetable wastes	1.0–6.0	1.2–2.9
Straw	2.5–5.0	0.4–1.2
Catch crops	2.0–7.0	1.3–4.6
Arable main crops (annual)	8.0–20.0	1.7–3.6
Perennial energy crops	17.5–25.0	0.8–1.3
Natural vegetation	0.5–15.0	0.05–0.2

[6]

Wood and wood wastes

A Firewood

Wood is *the* traditional fuel and even to-day perhaps half the wood harvested in the world is burnt as fuel. Even though this represents a huge quantity it is not a major contributor of energy to developed societies. Though it is used in industry, it tends to be restricted to special circumstances and situations and most of the wood is burnt in small-scale, dispersed and domestic situations for basic heating and cooking for the greater part of the population of developing countries. Though there are many wood-burning boilers, particularly in forest industries, and even some wood-burning locomotives in use in developed countries, these tend to be self-contained energy systems for specific applications.

Prime quality wood is a valuable commodity for construction, fabrication, the making of paper and various particle boards. Whenever it occurs in commercially exploitable quantities for these purposes it will not be used for fuel so that firewood, as used, consists of small lots of wood waste of various sorts including poor quality wood. It is wood waste, or rather wood residues, that constitutes the considerable potential biomass energy source that is considered here. This may be burned directly in all its forms, or like other biomass stock, be converted by thermal and chemical processes to gaseous or liquid fuel.

B Wood

The biomass of a typical commercially harvested tree is distributed in the following manner:

	(%)
Trunk	60–65
Top	5
Leaves and branches	10–15
Stump	5–10
Roots	10

Normally the last four categories are left in the forest as waste and more waste is generated in the processing of the trunk. Wood waste as a resource, as distinct from randomly collected firewood, arises mainly in three ways:

(*i*) as forest harvest waste. This can represent up to 40 percent of the total above-ground biomass of a clear-felled forest. In addition to this are the roots, which, though not generally exploited, could be considered as an available resource as methods of 'whole-tree harvesting' are developed.

(*ii*) as process plant waste. This is composed of sawdust, bark and trimmings from sawmills, pulp plants and factories and can represent up to 30 percent of the wood delivered to the plant.

(*iii*) as cull trees, being rough, rotten and dead trees and other material removed from timber plantations as part of their management process.

Further down the use cycle wood can form a significant proportion of municipal solid waste (MSW) but in general this is not separable from the rest of this material and cannot be considered as a separate resource.

Wood is composed primarily of cellulose and lignin, with various chemical extractives and, after analysis, residual mineral ash. A typical pine will contain approximately 50 percent cellulose, 20 percent hemi-cellulose and 30 percent lignin. Table I shows the proportions of these for common tree species in North America. Green wood also usually contains between 40 and 60 percent water which on air-drying typically falls to 15–25 percent. The commonly used oven-dry ton (odt) has 8 percent moisture.

Typically the gross energy content of wood is as follows:

Green wood (50% moisture)	10.5 GJ/t
Air-dry (20% moisture)	16.3 GJ/t
Oven-dry (8% moisture)	19.8 GJ/t

Table I Chemical composition of tree species (%). (Quoted by Molton *et al.*, (1978)

	Cellulose	Lignin	Pentosan	Anhydride	Residues
Trembling Aspen	56.6	16.3	18.9	3.28	4.4
Beech	45.8	22.1	21.0	4.76	6.34
White Birch	44.5	18.9	25.5	4.63	6.47
Red Maple	44.7	24.0	20.2	3.47	7.63
Jack Pine	45.0	28.6	11.4	3.93	11.07
White Spruce	48.5	27.1	10.6	3.6	10.2

Table II gives some examples for typical tree species in the United States.

Table II Energy values of common woods. (USDA figures)

US Species	Btu/lb (oven-dry)	GJ/odt
Ash	8928	20.7
Birch	9508	22.09
Poplar	5580–8600	12.96–19.9
Sycamore	5800	13.4
Pine	5491	12.75

C Wood waste

Though large residues are generated by forestry activity only a proportion is ever likely to be available for fuel use. There is a competing demand for various manufacturing purposes, and this demand is likely to grow with increasing world demand for wood products. For instance Grantham (1977) indicates that in 1970 harvested forestry residues in the United States totalled some 65 million oven-dry tonnes, the destinations of which were as follows.

	Mt(od)	(%)
By-product use (chipboard etc.)	21	32.3
Fuel use in plant	6	9.2
Fuel use elsewhere	12	18.4
Chemical	26	40.0
Total	65	

The increasing use of residues is indicated by only 23 Mt(od) remaining unused in 1977. However, there are possibilities for an overall increase in residues if improved forest harvesting methods are introduced including whole-tree harvesting and chipping at site which will involve the trimmings. However, the main impetus to develop these methods is the demand for paper and chipboard products.

Alternatively Zerbe (1977) points to 19 Mt(od) of forest industry process wood waste and bark being unused in the United States in 1977, and in addition to this, 120 Mt(od) left at the logging sites. Further, he identifies a new resource of 1000 Mt(od) of 'non-commercial' timber. This is rough and rotten wood, wood of unsatisfactory species and the like, that is largely to be found in the unmanaged timber stand of the United States.

Like the United States, Canada is also a country with very large timber resources, much of which are considered to be under-utilised. Over 32 percent of the land area is forest and it has been calculated that the maximum productivity from this, if fully managed and assuming an average annual growth increment of

6 t/ha/yr(od) would give 400×10^6 odt/yr with an energy content of 8×10^{18} J (8 exojoules) (Love and Overend, 1978). This is equivalent to the total Canadian basic energy demand in 1974.

In fact the current harvest (1974 figures) for all purposes is only about 51×10^6 odt with an energy content of 1 EJ. About 72×10^6 odt are direct fuel wood harvest, mostly used for small scale heating purposes. Table III gives figures for waste potentially available for additional energy use. Some of this is used already as about one-third of the energy requirements of the forest industry in British Colombia are supplied by burning waste.

Table III Quantities of wood waste potentially available in Canada per annum. (After Love and Overend, 1978)

	10^6 odt	Gross energy content 10^{18} J/10^{15} Btu*
Mill residues	7.5	0.14
Residues from forest operations	31.0	0.58
Unutilised timber in currently logged areas (cull)	20.0	0.37
Potential cull from currently unmanaged areas	52.0	0.97
Total	*109.5* (mt/yr)	*2.06* (quads/yr)

* Approximate equivalent

The availability of such large resources in North America is giving rise to a great deal of research into their use with a considerable swing, particularly by the forest and timber industries, to the use of wood wastes. In Canada serious study has been made of the large-scale production of methanol from wood as a petroleum supplement. Though there are considerable problems in the use of methanol or methanol additives to gasoline the technology is probably relatively easy and is being worked out on a large scale in Brazil with their alternative ethanol from sugar strategy. In Brazil also, their large timber resources and particular problems with petroleum supply are causing them to examine large-scale methanol as well as ethanol production.

Not all countries are as well off for timber resources as the United States and Canada. The countries of the EEC import 60 percent of their timber and the United Kingdom 90 percent. The timber import bill is in fact the next largest single commodity bill after that for oil for the EEC. In the face of increasing demand, even with some of the large-scale afforestation programmes proposed, the EEC could not become self-supporting in timber in the next few decades.

WOOD AND WOOD WASTES

The total wood consumption by the European Community in 1978 was of the order of 195 Mm³. With an average wood density of 2.2 m³ per tonne, this gives 88.6 Mt. About 75 percent of this is processed wood and the remainder pulp wood. Of the various operations, processing wood produces about 50 percent waste material and paper-making from pulp about 10 percent. Wastes from the overall quantities therefore amount to 33.2 Mt and 2.15 Mt. In addition to this, if a further 15 percent of the 78 Mm³/35.4 Mt Community harvest can be recovered as harvest waste the total waste potentially available as fuel will be 40.65 Mt. With an energy content of 16 GJ/t this will have a total energy content of 650 $\times 10^6$ GJ or 14.7 million tonnes of crude oil equivalent (Mtoe) — about 5 percent of the 1978 total Community energy demand.

Even when wastes and residues are collected, there is some doubt as to how much of them will be available for fuel use. We have seen that in 1970 nearly 10 percent of residues were already being used for fuel in forest industry. In Britain, about one third of sawmill residues are similarly used and there are movements towards increasing use prompted by increasing costs of other fuel. Looking at overall forest industry requirements it would appear quite likely that the forest and woodworking industries could absorb all their current wastes for fuel, plus most of the additional resource from more efficient management and harvesting.

However, for much of the material the fuel value may remain less than its

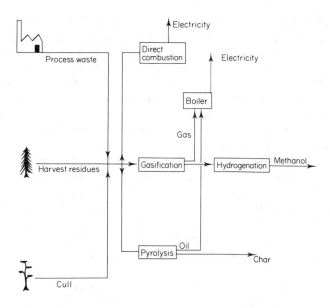

Fig. 1 Main routes for wood waste as fuel

value as industrial raw material, for board and paper products. If its fuel value remains less than that of alternative conventional fuels also, the tendency will be towards its use in products rather than for its energy content in the first place. If this were the case the general rise in demand for forest products indicates that these could absorb all available residues by the turn of the century.

The extent to which either of these paths will be followed will depend upon the developing supply and demand patterns for wood products and energy. Whatever transpires, the variety of local stock and conditions however, gives scope for a range of developments of wood for energy. These include direct combusion and thermal and chemical treatments to produce other fuels.

D Direct combustion

Wood may be burned as it is or be converted to other solid fuels like charcoal or wood pellets prior to direct burning. The residual char from pyrolysis treatments is also usable as fuel. The main problems with raw wood fuel are its high and variable water content and its variation in size which give rise to uneven burning and relatively low temperature. Domestic burning aside, as a fuel for large-scale heat generation it therefore needs drying and chopping or 'hogging' to a constant size for easy handling. A typical 'hog fuel' will be reduced to 5 cm pieces and have a moisture content of less than 40 percent.

The energy recovery route is via steam to be used directly in industrial processing or to drive electric generators. Wood burning at site is most viable if both steam and electricity is required, as it is for pulp and paper plants. Wood boilers have been in use for a very long time and are basically similar to coal-fired boilers. Many are dual coal/wood or multi-fuel types. Boiler efficiencies vary from 45 to 75 percent depending on the moisture content of the fuel. Typically they will be at 70 percent with a steam cycle efficiency of 40 percent and an auxiliary power requirement of about 10 percent, the overall efficiency of electricity generation from wood is usually taken to be about 25 percent.

There are many new types of boiler being developed as well as old boilers being adapted for wood fuel. In some cases combined furnace/pyrolysis units are being installed and moving bed systems are being introduced to improve efficiency. Specially designed boilers can cope easily with up to 50 percent moisture content in the fuel, the essential approach being to use a deeper combustion bed with wetter material. New designs using the forced vortex principle involving the injection of compressed air can achieve up to 90 percent combustion efficiency.

In the United States the old-style wigwam waste disposal burners used by the forest industries are being phased out due to pressure against pollution and energy waste, and these are being replaced by steam boilers. A notable example of non-forest industry use of wood is in the Eugene City power system in Oregon where there is co-generation of steam and electricity from hog fuel, the steam being used for heating and cooling. This has been in operation since 1951. Oregon is a heavily forested area of course, and up to date the economic use of wood waste has depended on low-cost fuel. In the forest industry, the largest types of boiler take up to 225 t/h of fuel. Work is going forward in the USA on the design of wood burning plants to produce up to 50 MW for general utility use.

Zerbe (1977) calculates that a pulp plant using the usual Kraft process, turning out 1000 t/day of pulp and needing 22.4×10^6 Btu/t (23.6 GJ/t) input of energy could obtain this from 4300 t of residual wood fuel with 50 percent water content, obtainable from its own logging area. Alternatively, following the 1973 oil price increase it was shown from studies in the United States that grid electricity could be generated economically from wood-burning power stations using between 100 and 1000 dry t/day wood, provided the transport costs of the fuel were low and the wood itself was non-commercial — in fact the product of improved management of previously 'stagnated' timber stands. This of course is only valid for certain types of locality and opportunity. Additional benefits of such schemes would arise from the increased value of the timber stands and the possibilities for the use of the process steam.

Charcoal is the oldest processed fuel in the world. It is produced by controlled burning or heating of wood in the absence of air, the effect of this being to prevent oxidation of the carbon compounds and their loss as carbon dioxide and monoxide. The result is almost pure carbon which has a much greater energy density, 29 GJ/t compared to say 15–20 GJ/dt for wood. The conversion conserves about 30–50 percent of the energy content of the wood which is generally higher than that reclaimed from wood by direct burning. It contains no moisture and burns at much higher temperatures enabling it to be used in the first instance for melting metals, and it was the main fuel for these processes until the general use of coal. A name sometimes used for it is 'biocoal'. Today there is little interest in it *per se* as an industrial fuel, partly because it is a product of relatively scarce wood resources. However, because of its convenience and greater ease of handling it is considered of interest for domestic and small-scale industrial heating in countries with ample wood resources. Its use as briquettes in solid fuel burners or slurry reinforced by small quantities of oil in oil burners requires little capital investment.

It is also safer to handle and use. It is not highly inflamable or explosive and if burnt at less than 600 °C less than 10 percent of toxic carbon monoxide is generated and no nitrogen oxides. Figure 2 illustrates a hot air burner/blower used for industrial heating purposes.

Charcoal as such is unlikely to become a major fuel, pelletised or briquetted wood waste on the other hand has a number of advantages that could make it an important solid fuel or fuel supplement.

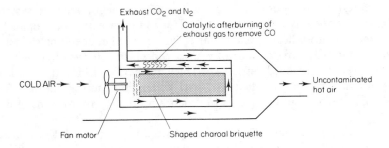

Fig. 2 Principles of a charcoal air heater. (After Thorensen, 1978)

Pelletisation of wood waste is an improvement on 'hogging' giving opportunities to:
  (i) reduce the moisture content,
  (ii) increase the bulk density,
  (iii) increase the heating value with additives,
  (iv) improve the handling and transportation characteristics.

Pelletisation is carried out on a number of materials including fertilisers and animal foods as well as municipal solid waste, wood chips and sawdust. A number of equipments are available for this. The process with wood waste involves size reduction, moisture reduction, compaction and addition of a binding agent. For fuel pellets a number of binding agents have been tried including bitumens, resins and gums. Some woods with high resin contents bind naturally if heated under pressure but this gives only a weak bond, as do most of the additives.

A new process in the United States (Terry, 1978) uses thermoplastic resin as a binding agent. These 'Franjon' pellets are much stronger than previous types and consequently easier to handle, transport and feed into boilers. The addition of the plastic material also considerably increases the energy content of the material — it is claimed from 8000 Btu/lb for raw wood, to at least 18 000 Btu/lb for pellets. These burn at higher temperatures than untreated wood, resulting in more complete burning and less residual ash. The thermoplastic used can be obtained from waste.

E Pyrolysis and gasification

Heat treatment of wood for various purposes is a long-standing technology. The production of charcoal by burning wood in a restricted supply of air to prevent full combustion is a pyrolytic process, though in this case the energy-rich gases are lost. Prior to the development of the petrochemical industry in the 1920s wood distillation was a major source for a number of valuable hydrocarbon liquids like acetic acid and methyl alcohol (methanol). Kilns for wood treatment are still commercially available and are mostly used for relatively small quantities of factory waste.

The products of simple dry distillation of wood waste in a retort include a gas, a liquid and a soluble fraction. For English oak typical proportions (excluding the moisture content) would be:

	(%)
Wood gas	14.5
Pyroligneous acid and tar	63.0
Charcoal	22.5
Total	*99*

The pyroligneous acid is the water-soluble part of the liquid fraction (wood spirit) and the tars the insoluble part. Wood gas is composed of non-condensibles and is mostly carbon oxide and hydrogen. Both the liquids and the gas are combustible and are potential fuel or fuel feedstocks.

E1 WOOD GAS

Producer gas from heat treatment of wood waste differs from digester gas in that its main fuel energy component is hydrogen rather than methane. A typical composition, from gasification in air, omitting water vapour is:

	(%)
H_2	20
CO	25
CO_2	10
$C_x H_4$	3
Higher hydrocarbons	1
N_2	40
Other	1
Total	*100*

This would have an energy content of about 150 Btu/scf (approximately 6 MJ/m^3). Table IV compares another version of wood gas to natural gas composition and properties.

Though there is some variation, in general the composition of the gas is largely independent of the composition of the feedstock because the end product is a mixture of simple gases. In some cases if there is a significant sulphur content in the feedstock hydrogen sulphide (H_2S) may be generated, but this can be easily removed by passing the gas through water. Most producer gas is in the range of 100–240 Btu/scf (or 4–8 MJ/m^3). However, if the air supply is

Table IV Composition of wood gas and natural gas. (After Vos, 1977)

Dry volume (%)	Wood gas	Natural gas
Hydrogen	10	–
Carbon monoxide	30	–
Carbon dioxide	6	0.2
Methane	1	96.0
Ethane	–	3.0
Nitrogen	50	0.8
Tar and oil vapours	3	–
Btu/scf	200	1026
flammable limits in air (%)		
Lower	12	4.8
Upper	74	13.5

replaced by pure oxygen as it is in the PUROX process the nitrogen content is eliminated and the energy density can go up to 11 to 19 MJ/m^3. Another route to producing the higher Btu gas is by hydrogasification where hydrogen is added to react with the carbon monoxide to give more hydrocarbons.

The principles of a simple gasifier are that the wood fuel is fed through the top of the combustion chamber via an air-lock system. The air (or oxygen) feed is at the base but the supply is restricted by the build up of a carbonised bed. This has to be periodically removed by moving a grate.

The conversion efficiency of low Btu gas generating units is typically 65–75% (for 100 Btu/scf/6 MJ/m^3 gas). A typical yield from a portable unit taking 54 odt/day is 10^5 m^3/day. This producer gas can be readily used with existing gas and oil burning boilers with modified burners. There are a number of wood gasification systems already on the market or under development in Europe and the United States. The largest operating systems can process up to 320 odt/day in plant being designed, for instance, by the Saskatchewan Power Corporation for a gas output with an energy content of 5×10^{12} Btu/h.

E2 METHANOL

The route of gas generation which involves hydrogasification of the producer gas leads to the production of quantities of methyl alcohol (methanol) by the catalytic reaction of hydrogen with carbon monoxide. Hydrogen can also be made to react directly with wood cellulose under high temperature and pressure to give a predominantly methanol liquid fuel with an energy content of about 30 GJ/t.

Major studies in Canada in particular have considered the possibility of using large scale unused timber resources and waste for 'gasohol' (methanol/gasoline mixtures) liquid fuel policy. As already mentioned Brazil is also considering large-scale methanol generation as well as the established ethanol programme, and large-scale plant has been procured from the German heavy chemicals industry.

WOOD AND WOOD WASTES

The technology for these strategies is available but the economics depend very much on comparison with alternative fuel costs. The price of methanol may for instance be forced down by methanol obtained from oil field gas, much of which is still flared off, and this could make the approach uneconomic.

E3 PYROLYSIS OIL

Wood oil is the non-soluble part of the condensate from wood pyrolysis. A larger fraction of this can be obtained by treating wood at higher temperatures and pressures than for gasification and replacing air with hydrogen or steam in the reaction. Wood can be produced in various types of pyrolysis plant that can equally well take municipal solid waste or crop residues.

The products are light oils with relatively low energy contents, notably low in sulphur. As such they can be conveniently used with mineral heating oils of higher sulphur content to reduce overall pollution levels. Alternatively they can be distilled to form more convenient forms of fuel. Potentially yields of between 100 to 200 kg per odt of feedstock can be produced with an energy content up to 30 MJ/kg. Table V compares wood oils from a 50 t/day field demonstration plant with typical fuel oils.

Table V Wood oils and fuel oils. (After Knight et al., 1977)

Percent	Wood oils		Fuel oils	
	1	2	3	4
Carbon	51.2	65.6	86.1	87.0
Hydrogen	7.6	7.8	13.2	11.7
Nitrogen	0.8	0.9	–	–
Sulphur	<0.01	<0.01	0.6–0.8	0.9–2.3
Water	14.0	10.4	–	2.0
Density	1.142	1.108	0.851	0.960
Flash point (°C)	111.6	1115.5	37.7	65.5
Energy content (MJ/kg)	21.1	24.5	45.5	43.1

F Other processes

Wood waste, or rather the sugar content in it can also be digested to produce methane. It is not however an ideal digester feedstock because of the relatively small sugar content. This can be increased by acid hydrolysis — for instance using sulphuric acid — to break down the cellulose, hemi-cellulose and liquid to sugars. In this case however, the material can be fermented to produce ethanol.

These routes for wood are not receiving as much interest as gasification and pyrolysis to produce fuel.

G Development considerations

Certain countries clearly have very large quantities of, at present, unutilised timber, forest and forest industry waste which could have considerable impact on their future energy requirements. Most countries have useful quantities that could be of significant benefit at least to their timber industries. The technologies for using them are available but their economic viability, as is the case with most other biomass energy strategies, depends primarily on the costs of competitive conventional fuels. Key factors are the low cost of recovering the waste and transportation cost. It thus makes most sense for the needs of the forest industries to be considered first.

Most commentators agree that all readily recovered waste being generated at present could be absorbed by the industry either for energy supply or for additional raw material for processing to chipboard and fibreboard. All prime timber is likely to find a ready market for the foreseeable future. Any 'unutilised stock' or waste in any quantity is only likely to arise by new methods and efforts in plantation management and harvesting, like whole tree harvesting. The economics of this are only beginning to be examined and most practical schemes underway aim at harvesting these untouched resources for processing rather than energy generation. A factor that has to be considered if a strategy of removing more of the residues is followed is the depletion of nutrients in the forest eco-system. If most of the normally recycled nutrients are removed they will have to be replaced by artificial fertilisation which could be an expensive business and an important addition to the debit side of the overall energy economy of the system.

The value of so-called wastes as energy feedstock must thus be equated with their alternative value as a process resource. They may also emerge to be more valuable as non-fuel chemical feedstocks. A number of pulp plants in the United States already use their processed wastes to produce other chemicals and beverage quality alcohol. These have, of course, already been chemically treated and wood is a less attractive feedstock when raw. The most likely direction of utilising wood for its energy content will be by the diversificaton of the forest industries to enable them to direct the basic resource material along whichever routes prove the most profitable.

[7]
Short rotation forestry–SRF

Short rotation forestry is a version of the energy plantation using woody perennial tree crops, the idea being that forest plantations are grown for their energy rather than their fibre content value. This entails a much more systematic approach than the virtual unmanaged firewood reserves of some countries. Though there are connections with normal forestry management practice some aspects are novel and overall the management style could be closer to that of large-scale plantation agriculture. At the time of writing no such schemes have been established though a number of detailed feasibility studies have been undertaken and considerable work on crop yields in experimental plots and associated propagation and plant breeding trials is being carried out.

A Why short rotation?

The over-riding reason that short rotation is essential in a forest energy plantation is the need for a sufficiently high return on the investment in land. A normal forest harvest cycle is 30 to 80 years and the total growth averaged over this time does not yield enough biomass to approach an economic return on its use as fuel rather than wood and fibre. In fact the estimated cost of the fuel value decreases with the shortening of the harvest cycle, as the cost of invested capital increases with time. The idea, then, is to use trees that grow very fast in the first few years, and crop them frequently — at intervals, if possible, of less than 5 years. The type of tree crop most suitable for this treatment are certain fast-growing hardwoods including poplar and eucalyptus species. These also have the added advantage of being able to sprout from their cut stumps thus effecting considerable saving in the cost of management by obviating the need for replanting after each harvest.

Cropping of tree plantations has a history in providing poles for construction and the cutting of willow wands for basket making. The willow woods on many of the islands in the river Thames are relics of this, now grown up after the practice has been discontinued. More recently it has been considered as a technique for producing paper pulp but it has not yet been proved to be satisfactory for this purpose. The idea of SRF as a version of energy plantations began to be examined systematically in the early 1970s as one of the wide-ranging responses to the energy crisis, notably by Szego and Kemp (1973). Biomass derived from SRF can be used in the same variety of ways as wood waste and the general advantage of fuels derived from SRF are seen to be:

They will be largely sulphur-free

The feedstock is renewable

The production of the raw material will not drastically disturb the land

Their production and use will not adversely affect the global thermal or carbon dioxide balance

They are non-polluting

The residues are recyclable.

The advantages of SRF in particular are:

Higher yields per unit of land

Quicker return on investment

Improved harvesting efficiency through mechanisation

Higher labour productivity through mechanisation

The ability to incorporate new cultural practices and new genetic material quickly.

Conversely, the disadvantages would be:

Higher establishment or management costs, with the highest cost in the first year which would also have a substantial effect subsequently

Because the plantations are essentially monocultures they may be particularly subject to disease and infestation

Very large areas of land will need to be acquired and/or modified

Because of the necessity for large-scale mechanisation the size of plantations will be strictly defined by economic considerations linked to the actual equipment used.

B Yields

High yields are essential if SRF is to be economic. Though, as indicated by Schneider (1973) plants can theoretically convert received solar radiation at a maximum efficiency of 11 percent, even under ideal field conditions achieved results fall far short of this and conversion efficiencies of forest plantations do not appear even to reach 1 per cent. Even a common level of 0.4 percent by high yielding field crops has been achieved by only a few tree crops.

In northern European conditions the generally accepted value is around 0.15 per cent conversion efficiency, equivalent to a dry biomass production of 0.3 kg/m^2/yr or 3 dt/ha/yr. The difference between this and the over 200 t/ha/yr yield apparently theoretically possible, suggests there is plenty of scope for increasing yields by cultivation practice and plant breeding. However, even the most optimistic projections based on experimental work fall far short of the maximum. Szego and Kemp (1973) suggest that at 0.7 percent conversion efficiency (equivalent to about 14 dt/ha/yr) could be achievable and would be sufficient to make the basic proposition economic. However, the possibility of achieving such yields at the very large scale envisaged has yet to be proved.

Many factors of climate, availability of water and nutrients, and plant vigour conspire to reduce the theoretical yields. Under field conditions in the Netherlands, Frissel *et al.* (1978) has shown that the optimum practical yield from fast growing poplars growing in ideal conditions is between 15 and 18 dt/ha/yr compared to a theoretical maximum of 44 dt/ha/yr. These correspond to conversion efficiencies of 0.7 to 0.8 percent and about 2 percent.

Many claims for different species yields exceed the optimum practical field values of 15–18 dt/ha/yr indicated by Frissel *et al.* for poplars. These can be accounted for by more productive species and growing in better climatic conditions. However,, some claims may be misleading due to imprecise definition of the measured unit in terms of its water content. The following terms are generally accepted:

Bone-dry = 0% moisture. (Here written as d)
Oven-dry = 8% moisture (od)
Air-dry = Around 20% moisture, variable. (What is usually meant by 'dry')
Green = Around 50% moisture, variable

Which of these is used in yield claims is not always stated.

Normal forest yields of 3 to 8 dt/ha/yr averaged over 30 to 80 years compares to quite modest projections of 12 to 25 dt/ha/yr over 4 to 10 years for coppiced SRF suggested by Gibson (1978) or 25–30 dt/ha/yr suggested by Inman (1977) as necessary for economic feasibility.

The kind of yield required can be compared to reported yields for a variety of species thought to be capable of coppice cultivation in Table I. Table II gives figures for eucalyptus species and different locations identified as suitable for SRF in a study reported by Eliseo and Marians (1978).

The energy equivalent for SRF biomass will by typically 20 GJ/t (dry) or

Table I Biomass yields from potential SRF tree types (North America). (After Inman, 1977)

	Location	Yield* (dt/ha/yr)
American sycamore	Georgia	10–25
Red alder	British Colombia	37
Hybrid poplar	Pennsylvania	10–22
Hybrid poplar	Winsconsin	17

*If 'air-dry' is meant here the fully dry yields will be 20 percent less

Table II Stem yields of eucalyptus species. (After Eliseo and Mariani, 1978 quoting Sieman, 1975)

	Rotation-years	Location	Green wood yield (t/ha/yr)	dt/ha/yr*
E. globulus	15	India	32.5	16.25
E. globulus	6	Italy	20.5	10.25
E. globulus	10	Portugal	15	7.5
E. regnans	24	Australia	22.6	11.3
E. regnans	9	Australia	11.0	5.5
E. grandis	8	Australia	16.0	8
E. grandis	10–15	Australia	8.5	4.25

*Assuming 50 percent moisture

16 GJ/t (air-dry). This represents both a low energy density in the fuel material, with a low spatial density. This means that very large areas of land are required to provide energy in significant quantities. Calculations show that with an average annual yield of 10 dt/ha generating 20 000 kWh of electricity at a thermal conversion rate of 35 percent, it will require over 65 000 ha of land to supply a 150 MW power station in continuous generation.

SRF schemes are thus only viable in countries with plenty of spare land.

Inman (1977) estimates that if 10 percent of the 108 million ha or so of forest, pasture and range land in the United States were converted to SRF-type energy plantations with yield of 25–30 dt/ha/yr yields, they could supply 4 Quads of energy — 4×10^{15} Btu or about 91 Mtoe.

Such areas of land are clearly very large and the closest parallels in scale are pulp mill requirements. A 1000 tonne a day pulp mill requires about 800 to 900 km^2 of forest land to keep itself supplied with raw material. With a solar energy conversion rate of 0.4 percent, this same forest area could supply enough fuel, which converted at 34 percent thermal efficiency could supply a 400 MW electricity generating station operating at an operating load factor of 55 percent.

Table III SRF area required for a 450 MW generating station.* (After Szego and Kemp, 1973)

Insolation rate (Btu/ft^2/day)	Insolation rate (MJ/m^2/day)	Area required in km^2 for solar energy conversion (%)		
		0.4	0.7	1.0
1200	13.6	1036	595	414
1300	14.75	958	544	388
1400	15.9	880	492	362

*Assuming a thermal conversion efficiency of 34% and 55 percent load factor

This is an average load factor for the United States and the electricity supply would be sufficient for about 200 000 people.

Variations on this are illustrated in Table III using different insolation rates and solar energy conversion factors. This shows that there is apparently a considerable advantage in more southerly latitudes, in the case of North America favouring the south east where there is also sufficient rainfall to support the tree crops. It also illustrates a very much greater advantage if much higher than current yields could be achieved through high solar conversion efficiency.

C SRF in practice

The main elements of a short station forestry energy plantation are illustrated in Fig. 1. The idea by Szego and Kemp (1973) has been developed in studies by

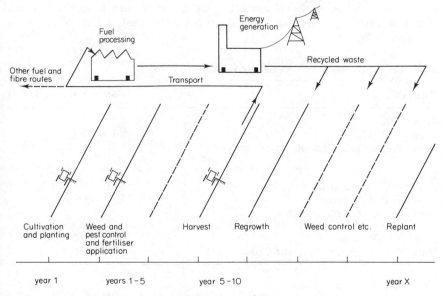

Fig. 1 Elements of a short rotation forestry scheme

Intertech Corporation in the United States (Fraser et al., 1976) . The chief features determined for a feasible scheme were:
Clone propogation of coppiceable species.
A clone planting density of between 9000 to 27 000 per hectare.
A first harvest at two years.
Repeat harvests at 2 to 4 years cycle.
Harvesting by agricultural type equipment similar to a 4-row silage harvester.

It was calculated that to justify the cost of mechanisation, allowing for a yield of 22 odt/ha, would require a scheme size of 11 500 ha. This would be divided into 4 units and would employ 86 full-time and 48 part-time staff.

The total initial investment for this (in 1975) would have been $11.4 million and the output of 1 million odt/yr would cost between $13 and $16 per odt. This would provide energy with a heating value, for comparison with other fuels at between $1.0 and 1.2 per million Btu (or per GJ).

A similar feasibility study carried out by Mitre Corporation for the United States Energy Research and Development Administration (ERDA) in 1976–77 came up with a cost of $20 to $30 per odt with a heat energy cost of $1.2 to $2 per MBtu/GJ, though there were possibilities of reducing this to $16 to $20 odt and $1.0 to $1.5 MBtu/GJ.

These fuel/energy costs can be compared with prices for coal and natural gas in the United States at the same time in Table IV. Though the prices are clearly comparable to the higher Btu fuels it can be seen from the low Btu coal that energy density (and also convenience of use) is a major factor in practice.

Table IV Comparison of fuel-energy values, 1975/76

Fuel energy	Cost per MBtu/(1 GJ)
Natural Gas at *ca* 1000 Btu/Scf (from average prices for industrial users)	$0.7–$2.2
Coal at 11 000–13 000 Btu/t at 7000–10 000 Btu/t (from average contract prices)	$1.0–$1.2 $0.24–$0.59
Proposed SRF Fuel (Intertech and Mitre studies at 5700 Btu/t)	$1.0–$2.0

C1 GENERAL FACTORS

Water, either as rainfall or irrigation is of key importance in obtaining the sort of yield required. Fraser *et al.* (1976) claim that a minimum of 50 cm of rainfall is necessary to give a 'minimum acceptable yield' of 17 odt/ha/yr for economic SRF production. Most other commentators indicate that a rainfall of 60 cm is essential before non-irrigated SRF can be considered. This, of course, imposes geographic limits or demands added investment in irrigation.

Generally second-class rather than first-class agricultural land is proposed for SRF, land that may be under forest or scrub already. However, special types of land like wetland, provided it is of sufficient extent, is considered provided there are tree species that will grow on it. The main requirement is for suitable soils of sufficient depth and long slopes of less than 30 percent to allow mechanisation.

Vegetative propagation is preferred to seeding as propagation in bulk is cheaper by this method and much greater control can be exercised over variety, maintaining high-yielding stocks with other desirable characteristics like rooting ability and disease resistance, and eliminating the diversity generated by sexual reproduction. Potentially, cloning from tissue fragments cultured and treated with hormones for appropriate development — like encouragement of rooting — could give rise to millions of plants from one parent.

C2 ENERGY BALANCE

Obviously, to be worthwhile, considerably more energy must be obtained from an SRF plantation than goes into its establishment, running and harvesting. Calculations for the Mitre Study claim an energy input balance of 10:1 to 15:1. However, calculations by Siren (1976) for Swedish conditions point to lower ratios. Detailed calculations of all inputs, including the energy content of the fertiliser and machinery* add up to 24 400 kWh/ha/yr, see Table V. This is the energy content in 12 m^3 of dry wood which, with an average density of 0.4 t/m^3, means 4.8 dt. Therefore to gain any energy, more than 4.8 dt/ha/yr have to be produced and with practical yield of possibly only 15–18 dt/ha/yr (see Frissel et al., 1978), ratios must be only 3 or 4:1. To get better than this energy inputs must be reduced.

Table V SRF plantation Sweden energy input equivalents

Management factors	Energy input per ha/yr in kW/h
Production of cuttings (40 000/ha) plus transport	400
Drainage, soil preparation, establishment	2000
Fertiliser and application	16 000
Control of water regime and silviculture	700
Harvesting operations	700
Processing and drying crops	400
Energy input in manufacture of machines	2000
Energy loss due to production failure– assume 10%	2200
Total	*24 000*

C3 LAND AVAILABLE

Even when obtaining the highest yields proposed, the practical realisation of an SRF plantation for any significantly large production of energy requires large

---

*This is the energy input in the manufacture of the fertiliser and in the machinery in use on the site in proportion to its use and life

land areas, the order of size of which has already been indicated. However, these areas are not as large as those required by natural forests to supply the same quantities of energy feedstock.

In the United States it is estimated by Kemp and Szego (1974) that the total energy requirement of the country could be met by energy plantations covering some 65 million hectares converting solar energy at 0.4 percent efficiency. This is in fact only one third of the land considered suitable for this purpose that is presently unmanaged forestry, range and scrubland, enjoying at present only marginal use. The allocation of such large areas even of marginal land would, of course, mean considerable political and organisational problems. Similar calculations could be made for a number of other very large countries. The idea is not valid for small densely populated countries and for areas short of water.

Sweden is an example of a smaller timber producing country without a high population density. Here, rather than using land already forested it has been suggested that up to one million hectares of poor land could be used for SRF including 250 000 ha of marginal agricultural land and 750 000 ha of low altitude and coastal marshland. This could supply energy equivalent to one third of the currently imported oil. Similarly it has been suggested for Norway (Thorensen, 1978) that as much as 4 million ha of marginal land could be available for this purpose.

Elsewhere in Europe much less land is available, certainly no reserve of prime agricultural land. However policies proposed by the EEC are to switch over 4 million hectares of marginal agricultural land in the Community to forestry as part of the Mansholt Plan and some of this could be considered for SRF. Up to 1 million hectares of this land is in the United Kingdom and Ireland, mostly hill land.

D Research and trials

All aspects of SRF are objects of study, with programmes in most cases dating back to the mid-1970s. The earliest work on this, however, goes back some twenty years earlier when the idea was examined in the southeastern United States for providing paper pulp. Studies being conducted now are for specific prospects, and species and areas rather than on the general concept. The Bioenergy Directory of 1979 lists 31 research programmes, mostly government funded, in the United States and Canada, involving universities, large companies like GEC and utility and power companies working on SRF related subjects.

Work on species selection covers a wide range of coppiceable species including sycamore, sweetgum, cottonwool, alder, locust, and maple and poplar hybrids. It has been determined that 95 of the 487 eucalyptus species could be suitable for SRF (Eliseo and Mariani, 1978). Of non-coppice species, pines are under examination and mesquite (*Prosopis* sp.). This occupies some 29 million hectares of scrubland in the southwestern United States which could be used for a modified form of energy plantations. Mesquite is of particular interest as, like locust, it is a member of the Leguminosae and has the ability to fix nitrogen from the air and is thus, in part, self-fertilising.

Selection and hybridisation work is combined in a number of cases with field trials or plots of up to 10 ha with the main interest in yields and overall energy balance. Cultural practices, including fertilisation trials, optimum spacing and harvest methods are being investigated. In one case trials are being conducted using municipal waste as a low cost fertiliser. Other work is being done on mass propagation of suitable species.

The US Department of Energy is planning to establish a pilot silvicultural biomass farm in South Carolina to carry out large scale trials of all aspects of SRF. Plans for this drawn up by Mitre Corporation specify a planting density of 6740 trees per hectare using sycamore (*Plantanus occidentalis*) European alder (*Alnus glutinosa*) and slash pine (*Pinus elliotti*). The plots will be drip irrigated and also fertilised by this method. Expected yields are 20 odt/ha/yr with a 6-year rotation.

In Europe an SRF prospects have formal part of an energy from biomass research programme funded by the European Commission and there is additional work in national programmes, notably in Sweden. Mostly this has been feasibility study work on potential yields, but some plot trial work is underway.

In general the work can draw on a solid background of work of forestry investigations and plant physiology. Ireland, however, has the prospect of having the first large-scale trial in Europe. Here the working out of large areas of peat land to fire power stations has left areas of land possibly suitable for SRF and with a complete infrastructure of roads, light railways and power stations already in place. Work on species plot trials and wood burning for electricity generation is laying the basis for a large-scale demonstration project. Of the European countries, Ireland is probably the most promising prospect for SRF. The climate is close to ideal with high rainfall and a low incidence of frosts. The population density is low and there are large areas of marginal land available. Current forestry is, however, very limited and there is a great deal of scope for expansion. Also with very limited resources of fossil fuels the national energy situation is weak and any prospects must be considered.

E Viability

In the first instance the viability of SRF depends on the achievement of a sufficiently high yield to give a large enough energy output/input ratio to carry the costs of the scheme and for the energy produced to be competitive in price with equivalent energy obtained from other sources.

On the first of these points there is room for considerable doubt about projected yields. In general plot trials do not seem to match up to the productivity assumptions made in the feasibility studies. The highest claims for yields and solar energy conversion efficiencies are particularly hard to verify from the literature. Further, they apply only to small-scale plot trials and the obtaining of the level of plot trials yields at field scale is notoriously difficult in agriculture. In forestry, over the very large areas necessary, it must be even more difficult. However, there is certainly great scope in plant breeding. In agriculture, wheat yields in England increased 84 percent between 1947 and 1975, and 50 percent of this increase has been attributed to new varieties. In comparison much less work has been done on high yielding tree species of the types considered for SRF.

Failure to achieve high enough yields could be compensated for in reduction in operating costs but another factor that arises in this case is that much larger areas of land will be needed to supply a certain tonnage of feedstock with consequent disadvantages.

A factor in favour of all alternative energy schemes like SRF is the rising cost of conventional fuels and their almost certain increase in cost for the foreseeable future. Though this may allow lower yields to be economically viable through their increased value, the advantage is partly offset by the increased cost of the energy going into the running of the scheme.

SRF is not opportunistic in concept like waste utilisation schemes and elements of subsidy only enter into it through the possibilitities of using urban or industrial waste as fertiliser, or by using low value land for which there is no more serious competitive use. Energy production is the primary purpose of the schemes proposed and not an add-on benefit to other schemes. However, it could be considered so if SRF were to be proved a satisfactory approach to fibre production — say for paper pulp. Alternatively, use of some of the fibre or crop extracts could be considered as add-on benefits to energy production.

In other words SRF has to stand on its own two feet and there are a number of major factors working against it in the internal economics of the schemes and from outside.

Internal economic factors, apart from that of yield, are:

(a) *The need for low cost land*
The fact that the land is low cost usually means that it has deficiencies with regard to agricultural or conventional forestry use. The need then, is to identify tree crop species giving high yields in the particular ecological conditions. As the need is for raw biomass rather than for specific bio-products this is probably not too much of a problem.

(b) *The development of cost-effective methods of propagation*
Work on cloning for use with coppiceable hardwoods shows promise here.

(c) *The development of methods of mass planting, cultivation, crop treatment and harvesting*
This must involve the development of new types of very large-sized traction machinery, large scale irrigation or water level control systems and aerial application techniques. All are costly and energy consuming.

(d) *The cost of additives, fertiliser and pesticides*
Fertilisation is essential for high yields and unknown disease and pest problems lie in wait for the very large-scale monocultures proposed in most schemes. Some alleviation of the fertiliser cost may be possible by using urban and industrial waste and the fuel residues can be recycled. The use of nitrogen fixing plants as or with the crop could be of some help.

(e) *The development of fuel processing techniques*
As shown in the chapter on wood wastes, proved techniques of processing and use of the material are available and they are not likely to be a constraint with the overall economics of the schemes.

External factors include the sheer sizes of the proposed schemes which must give rise to big political problems in land acquisition and assignment. These will be compounded by major social ecological impacts which have not yet been considered in any detail. The probability is that even in highly favoured countries the number of sites will be strictly limited.

Other factors are:

(f) *Competition for alternative use of the land*
This must vary considerably with the different circumstances in different countries. In Europe the chief competition for any 'marginal' land switched out of agriculture is conventional forestry. As the European Community imports about 60 percent of its timber needs, and the United Kingdom nearly 90 percent there are active policies for major afforestation that will obviously compete for land with SRF.

(g) *Alternative use of the product*
The question of using the SRF product for uses other than in providing energy has already been touched upon. Overall there is a big question mark over this subject as most of the factors for and against SRF apply whether the product is used for energy, fibre, or (in specialised circumstances) chemical feedstock, and if there is any competition relative price movements will decide the issue.

In short, the economic viability of SRF schemes hinges on satisfactory demonstration on a suitably large scale, but the ideas are sufficiently promising for a large demonstration to be undertaken. Should these prove the case the time-scale of schemes indicate that they will be unlikely to be widely implemented before the turn of the century.

[8]
Sewage and municipal wastes

These arise inevitably in any human environment and have long been considered, to a limited extent, as a source of energy. Sewage, which is essentially a liquid waste, and solid municipal and industrial wastes which are generally combined in collection, are generally streamed separately in any community. Because of this, as well as because of their differing nature, treatment and energy recovery methods differ considerably, as does the relative quantities of the materials. For instance the very low solids content of sewage means that it represents in any circumstances only a very small resource. However, it is handled in such a way that its use for energy production in the form of gas can be readily realised. It is, in fact, a main constituent of the inputs to the extensive Chinese and Indian domestic digester programmes and it has been generated for decades for use in the advanced sewage treatment plants of developed countries.

Solid wastes from households and industry is a much larger potential source of energy than sewage and because of its greater variations in composition and the variety of forms and situations in which it occurs, has more varied possibilities. In both cases they represent a problem to the community which is usually prepared to invest in methods for their disposal. Energy that can be extracted from them economically therefore represents cost savings. Conversely they are attractive for energy generation because much of the system of handling and treatment will have been paid for out of the disposal budget so that energy generation could be regarded as subsidised. Here then are perhaps, in the first instance, the most readily achievable energy from biomass prospects.

A Sewage

A1 NATURE AND QUANTITY OF ARISINGS

Sewage consists primarily of human excreta, admixed with variable proportions of industrial effluent. Its main component, therefore, comprises that very limited proportion of total biomass production which finds its way into the human stomach, minus that part of the ingested biomass which is consumed in human metabolism.

Callaghan et al. (1978) estimated total present biomass production in Britain at 181.1×10^6 t/yr, which is 47% of their estimate for potential production, and represents 7.5 t/ha/yr. This quantity of biomass amounts to 3.17 MTJ/yr, or 56.6 GJ/yr to each member of the population. Compared to this figure, the quantity eaten is quite small; according to Elton (1973), 13.3 MJ/head/day finds its way into food supplies (much of which is imported foods) and equates with 4.85 GJ/head/yr, 8.6% of indigenous biomass production. It is not easy to assess how much of this food is actually eaten, after wastage of various kinds has been allowed for, but the energy requirements or allowances for the whole population, taking account of the sex and age structure of the population, can be calculated from values tabulated by Sinclair and Hollingsworth (1969) and amounts to 9.17 MJ/head/day overall, a figure which extrapolates to 3.35 GJ/head/yr or 0.19 MTJ/yr for the population, or 6% of local biomass production. The amount consumed has to include the material that will be excreted, so the total ingested looks like about 10.56 MJ/head/day, 3.85 GJ/head/yr.

The proportion of this which is utilised in human metabolism may be judged from physiology, ingestion of a typical 2500 kcal/day adult diet resulting in the excretion of about 100 g dry matter and, on this basis, excretion by an average member of the population would be 87.6 g/day dry matter. His diet, with a 9.17 MJ intake, would typically contain 440 g of metabolisable organic solids; the excreta would contain 27.2 g inorganic solids and 60.4 g of organic solids; food intake would come to 527.6 g of dry solids/day. The excreted organic solids contain about 23 MJ/kg, and hence amount to 1.39 MJ/head/day, almost exactly 0.5 GJ/head/yr. This, then is the basic resource which comprises sewage, apart from the industrial effluent component, and in the UK it amounts to some 28 000 TJ/yr, 0.64 Mt of oil equivalent. It is tiny in national terms, as must be the case in all developed countries; in the UK it is about 0.3% of national energy requirements before conversion, about 0.18% after conversion to gross biogas, at 60% conversion efficiency.

Actual arisings of sludge frequently depart quite widely from the arisings as calculated above on the basis of physiology, for several reasons. The first is the variable proportion of industrial effluent, which is reported to commonly contribute from 10–25% of the solids and may be more in highly industrial areas. Secondly, cities are areas visited daily by a work force which may be largely resident outside its boundaries and hence the sewage solids arise out of proportion with the actual city population. These two factors lead to arisings in London and Manchester of well over 100 g of sludge/head of population/day, and even figures up to 200 g have been given for highly industrialised and densely populated areas. Correspondingly, dormitory suburb areas may show below average arisings. Thirdly, sludge arisings do depend upon the method of treatment of the sewage. Full treatment of sewage involves primary, secondary and tertiary treatment, but some works do not apply all these stages; the sludge

SEWAGE AND MUNICIPAL WASTES

output is then significantly below the rate of incoming solids entering the works in the raw sewage. Even with full treatment, there are losses to atmosphere of a few percent during the treatment stage and losses, mainly of inorganics, in the final effluent. Typically, these combined losses would be about 6% of the incoming material. Applying this to the per capita arisings calculated on the basis of physiology, the sludge arisings from an area with a non-mobile population of typical age and sex structure, with zero contribution from industrial effluent would be $87.6 \times 0.94 = 82.3$ g of sludge/head/day, a figure which is very close to the actual experience at many works, and equates to 30 kg/head/yr. The actual mean experience of all the water authorities in the UK is 27 kg/head/yr, according to the Department of Environment (1977), although this is a figure which necessarily encompasses the effects of disposing of (*i*) raw sewage by authorities discharging to the sea through pipelines, (*ii*) the addition of the industrial effluent contribution, (*iii*) the paper component, (*iv*) cases of incomplete treatment and (*v*) the proportion of the population not connected to any sewage system.

A2 METHODS OF SEWAGE TREATMENT

The primary treatment of the sewage is by sedimentation in tanks, in which all but the smallest suspended solids settle and oil and grease float, and can be skimmed off the surface. Retention time in these tanks is usually about 4 hours and the sludge is typically grey and slimy, with an offensive smell, and the sludge that accummulates usually contains just over 50% of the incoming solids. This is the primary sludge, those that are formed during the subsequent stages having rather different characteristics.

The second stage, where it is carried out, is an oxidation stage, and the two principal methods used are the trickling filters and the activated sludge process. The trickling filter consists of a bed of stones or other coarse material over which the sewage flows, distributed by slowly rotating arms equipped with nozzles and deflectors. Air is drawn in by temperature differential, keeping up a supply of air to the biological process. A slime, containing the active microorganisms, adheres to the stones, or to plastic filter media that are being increasingly used in their place. The principle of this system is that the organisms in the filter bed make use of the organic material in the waste in their own metabolism for the synthesis of cell constituents and for energy, thereby cleaning the waste. The organisms in the filter bed are, therefore, constantly on the increase, and the effluent from the bed contains a great many of them in suspension. They are much more easily removed from suspension, however, than the solids in the feed to the secondary treatment, and they are separated by passing the effluent from the process to a secondary sedimentation stage and allowing them to settle. This constitutes a secondary sludge and, if a third sedimentation stage is used, a tertiary sludge. In full treatment these sludges that separate after the primary stage may contain 40–45% of the original incoming solids. The actual trickling filter units are of many kinds, capable of different flow-rates and rates of BOD (Biological Oxygen Demand) removal, with different sizes and shapes of packing, different degrees of effluent and solids recirculation and various types of air circulation.

The second main type of oxidative treatment, an alternative to the trickling filter, is the activated sludge process. This originated from attempts to purify sewage by blowing air into it. This led, over a period of time, to a suspension of flocs of voraciously feeding organisms, which settle readily when aeration and agitation are stopped. In this process, as in the trickling filter, the floc organisms are feeding on the organic solids of the waste and, when the process has been sufficiently completed, they themselves can be collected and removed by sedimentation, purifying the waste stream. By adding fresh sewage to already treated waste containing the floc, high rates of purification can be achieved. The process, as practiced, therefore involves returning a proportion of the activated sludge to the aeration tank, discharging the balance of the sludge as a solid waste, aerating the sludge/sewage mixture to purify and, finally, settling the aeration tank effluent. The process is generally run continuously. The organisms involved in the process comprise a complex mixture of bacteria and protozoa; they have been reviewed by Pike and Curds (1971) who list 18 bacterial genera and 11 diverse protozoans. The aeration tanks may be 10–15 feet deep and detention time is between 4 and 8 hours. Aeration is absolutely necessary to supply oxygen at an adequate rate for the process, and this is done by means of diffusers, or by rotating paddles or brushes.

Whichever of these methods is adopted, the sewage solids are converted to a mixture of primary and secondary sludges requiring disposal. The secondary sludges from the two oxidative processes have lightly different characteristics; that from the trickling filter method, is known as 'humus' sludge; the sludge from both processes is brown, flocculant and relatively inoffensive with regard to smell, but the sludge from an activated sludge process can putrefy fairly quickly after being separated. Disposal is mainly by spreading or tipping on land (75% in the UK at the present time) of which about two thirds is disposed of usefully by spreading onto agricultural land and reclamation areas and one third is simply tipped. The remainder is mostly dumped at sea, with a tiny proportion (3%) being incinerated after air-drying. Land-spreading, tipping and disposal at sea all involve transportation, and in this matter the low solids content of the sludge is a great disadvantage, being only 3–6% solids and perhaps some 4.5% on average.

Sludge from the trickling filter process is commonly higher in solids (4–7%) than that from the activated sludge process (0.5–4%), with primary sludge being from 2–6%. The sludge for disposal is commonly mixed primary and secondary types and is usually thickened prior to disposal to save on transport; it then reaches 6–12% solids. This thickening is accomplished by mechanical dewatering, either by vacuum filtration or filter pressing.

A3 DIGESTION OF SEWAGE SLUDGE

In the past anaerobic digestion has been more frequently practiced at sewage works than elsewhere because, apart from the energy source generated, other advantages accrue to the operation of the works as a result of the process. Digestion reduces the amount of sludge by converting some of the solids to methane and hence reduces disposal costs. It also improves filtration characteristics of the solids, enabling them to be more readily de-watered after digestion; it also helps to destroy harmful pathogens and reduce odour. This means that the economics of sewage works digesters do not depend solely upon the value of fuel gas generated. Indeed, it seems likely that the lack of economic pressure upon the design and operation of digesters at sewage works may have been responsible for some lack of initiative in reducing costs to the full extent that can be seen to be possible by comparison, for example, with the types of farm digester now being installed.

These sewage works digesters may employ ambient temperature, and accept gas yields of about 500 l/kg solids from primary sludge, 350 l/kg from combined sludges, or alternatively, heat the digester, usually to about 30 °C to increase yields from primary sludge to about 750 l/kg, or 500 l/kg from mixed sludges. In the latter case up to about 40% of product gas may have to be employed for digester heating, with the proportion for this purpose varying from summer to winter, but, with a residence time of only 15 days compared with 30 to 60 days for a digester operating at ambient temperature, there is far better utilisation of the capital in the digestion plant. There is also a variety of two-stage versions of sewage digestion, and these usually represent an attempt, not only to maximise gas production, but also to facilitate a sludge sedimentation stage within the digester itself. As in the case of digesting agricultural residues, a scum may form on the surface of the digester contents and require breaking up by agitation, mechanically or by gas circulation, or by jets directed onto the scum layer.

The gas evolved usually has a thermal value of about 24 MJ/m^3. At 500 l/kg solids, the theoretical production level, based on per capita excreta output, would be 41 l/head/day, or about 1 MJ/head/day, 365 MJ/yr. This compares with the actual experience of sewage works of about 30–35 l/head/day, a mean of 0.78 MJ/day, 285 MJ/head/yr. This practically experienced rate of production represents 390 l/kg of solids, or 9.36 GJ/t of sludge solids, just under 60% conversion efficiency.

There appears to be little chance of sewage sludge gas from large sewage works entering the general energy economy because gas net production is generally less than that needed to run the sewage works. The power demand for this purpose is generally rather more than the 230 MJ/day of electricity, 2.7 kW, that may be generated, from the gas per 1000 of population, (it takes 1.0–1.5 kW per 1000 population to run an activated sludge plant alone) and this results in various degrees of self-sufficiency of power in sewage works that digest their sludge, ranging from about 30% to 90%, depending upon the type of sewage treatment process and the efficiency of the digesters and generators.

Gas production appears to be economic at all large sewage works, and at least 150 works in the UK use the process (Wheatley and Ader, 1979), but there is more difficulty with small ones. If neighbourhood or area plants for digestion of dairy waste or green plant matter become general, the sludge output from smaller sewage works could be carted to these and added as a supplementary feedstock subject, however, to rather strict monitoring and control of possible contaminants in the sewage, such as heavy metals, chlorinated hydrocarbons and synthetic detergents, which can inhibit the digestion process and damage agricultural land when the residues are spread.

A4 ECONOMIC CONSIDERATIONS

As a feed to anaerobic digestion, sewage sludge (350 l/kg) falls between dairy waste (250 l/kg) and pig waste (450 l/kg) with regard to its productivity. It would clearly be available as free feedstock from sewage plants without digesters of their own, but the difficulty would come in the matter of disposing of the residues. Some farmers, clearly, are prepared to allow sewage sludge on their land and have presumably satisfied themselves that contaminant levels are not serious in the case of the particular works from which the sludge is accepted. Such a farmer could accept any available undigested sludge and install his own digester to produce energy on the farm, though he would presumably be left with the task of spreading the residues himself. If the nutrient value of the residues was thought to justify the spreading costs, such gas might well be produced in an inexpensive on-farm digester run without resorting to additional farm labour, for about £1.50 to £2.00 per GJ of net gas production, which could be quite attractive. The sludge would similarly be economically attractive at a large area plant set up to digest dairy waste or crops, provided the obstacle of contamination could be overcome, because as free feedstocks offered with free delivery to site and a gas yield above that obtainable from dairy waste, the sludge gas could be produced much more cheaply than the main output. However, any consideration of using digester residues for animal feed would be inhibited. In such a case, siting the digester next to the sewage works could offer considerable advantage, since sludge could then be fed to the digester directly, without any need for drying or mechanical de-watering. The apparent disadvantage of a low solids content sludge would really be no problem here, since liquid would have to be added to either solid dairy waste or green plant material before digestion and, provided that the ratio of sludge solids to agricultural residue solids in the feedstock was low, the presence of sludge solids in the residues ex-digester would have little impact on the waste-handling arrangements of the digestion plant. In some instances the cost saving to the water authority could be very large; where, for example, disposal at present is by

SEWAGE AND MUNICIPAL WASTES

mechanical de-watering followed by disposal at sea, the costs incurred would be in the region of £25/t of sludge solids; at best, therefore, if disposal after treatment in the area digestion plant was free because, being admixed with a large excess of agricultural wastes, the sewage wastes caused no problems, the savings made would represent a credit of £2.67/GJ of gas produced. In the case of underwatered sludge at present being land-spread, the saving might be only half as much. Even if the sewage works concerned was large enough to contemplate its own digester, the collaboration and juxtaposition with the area digester would be worthwhile, to gain economies of scale, and to ensure the ability to run the sewage works completely on digester gas.

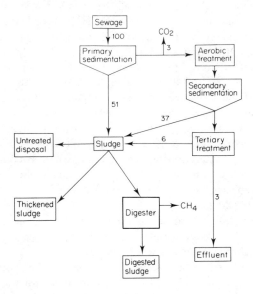

Fig. 1 Flow diagram of a typical sewage works process

B Municipal solid wastes – MSW

MSW is the familiar household and commercial garbage generated in every community, composed mainly of waste food, paper, plastic, rubber, metals, glass and ceramics. It varies in composition from time to time and place to place; for instance there tends to be less paper in European MSW than that in the United States, but the main component – in which the main possibilities for energy reclamation lie – is cellulosic substances.

Obviously the largest quantities of MSW are generated by the largest communities, and the more advanced countries. For instance MSW quantities in the

United States are estimated to be in the order of 200 Mt in 1980 (Greco, 1977), the calculation based on a figure of 1.3 to 2.2 kg per person per day. Compared to this the United Kingdom currently generates about 23 Mt/yr with London producing about 13 Mt of this or over 35 000 t/day. Production in the European Community of 9 countries was estimated at about 58 Mt in 1975 and is likely to rise at a rate of 5 percent per year to 95 Mt in 1985 (Europool, 1977). However, though these quantities represent a very substantial biomass resource, calculation of the energy that could be potentially recovered from this indicate that it can never become a major source of energy supply overall.

With such mixed material, highly variable from place to place and even from time to time, the energy contents naturally vary very considerably. High energy contents for MSW are usually related to high standards of living. For instance in Europe the average energy content of MSW is thought to be of the order of 10 GJ/t, while in the United States it may be as high as 14 GJ/t. On this basis it is possible to calculate that the production of the European Community of 94 Mt in 1985 would represent 940×10^6 GJ or 21.3 Mtoe. This would be between 1 and 2 percent of the projected total energy demand. However, even if all the MSW could be collected and processed, conversion losses would ensure that the energy recovered would be only a third to a half of this and in practice it could not supply more than 1 percent of the energy requirement. In some energy use sectors like electricity generation or space heating where it could be used most readily, the percentage contribution could be much higher. For instance in Germany it has been suggested that MSW could service between 15 to 20 percent of household heating demands.

Compositions of MSW for Europe and the United States are shown in Tables I and II. In these tables 'combustibles' for Europe equates with 'total organic material' for the United States. The distribution in the second table between digestible and non-digestible organic material separates the proportion that will generate methane-rich gas under anaerobic conditions and that material from which energy can only be recovered by burning or heat treatments (pyrolysis). This fraction includes significant amounts of synthetic plastics.

Typically, in a developed country, about 75 percent of MSW will be of organic origin and thus combustible, and most of this will also be organically digestible to produce methane. Only a small proportion however, is currently used to produce energy, though there is increasing interest both in alternative energy sources, and reclamation from wastes. Most material after some sorting for re-useable materials like rags, paper and metals, is dumped as land-fill or at sea. Where dumping sites are hard to come by or they are too distant and costly for material to be transported to them, some is incinerated. Like most things, these options are becoming increasingly expensive and energy recovery is seen as a method of defraying the expenses of waste disposal, rather than as a primary end in itself, and is usually combined with reclamation and recycling in integrated

Table I Typical composition of municipal solid waste in Europe. (After Schlesinger, 1977)

	Percent by weight
Combustibles	
Paper	35–60
Garden wastes	2–35
Food	2–8
Cloth	1–3
Plastics	1–2
Non-combustibles	
Metal	6–9
Glass	5–13
Dirt	1–5
Moisture	20–40

Table II Average composition of municipal solid waste in the United States. (After Wright et al., 1978)

	Percent	
Paper	42.0	
Food and garden waste	27.0	
Total digestible organic material		69
Plastics	2.0	
Textiles, leather, rubber, wood	4.0	
Total non-digestible organic material		6
Ferrous metal	8.0	
Aluminium	0.7	
Other metal	0.3	
Glass and ceramics	9.5	
Stone, dirt and other	6.5	
Total non-organic material		25
Total		*100*

systems. Removal of non-combustibles like metals and glass has the advantage of leaving a more organic-rich raw material for energy production, and the methods used may leave it in a more homogeneous and more easily utilisable state.

The basic problems of this feedstock are its variable composition and generally high moisture content — 30 to 50 percent. Difficulties of handling and transport are combined with relatively low energy contents of 14 to 18 GJ/t or 6–8000 Btu/lb. Though a number of energy recovery schemes have been in operation for years, to date little of the MSW stock is in fact used for this purpose. For instance of the huge US stock it is estimated in 1977 that only some 2 Mt/yr were being treated for heat recovery, (Schlesinger, 1977). Actual and potential methods for energy recovery, most of which are being experimented with are:

Direct combustion of unprocessed material, or treated material usually known as refuse-derived solid fuel — RDSF.

Anaerobic digestion to produce high Btu biogas

Pyrolysis by heat treatment to produce a lower Btu gas.

Direct combustion of MSW or RDSF usually produces low-grade heat, used either to raise steam for electricity generation, or process heat for manufacturing or space heating.

Biogas and pyrolysis gas may be burned directly to generate steam for electricity generation, or it may be fed into gas distribution grids. In the last case it has to be up-graded to the same Btu value as the pipeline gas by further processing. Potentially the gas can be further processed and energy-up-graded to give liquid fuel like methanol. Figure 2 shows some of the routes for energy recovery.

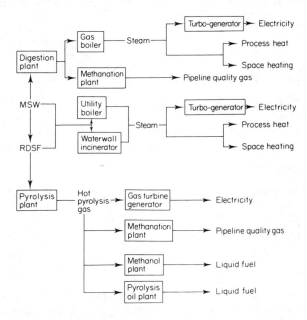

Fig. 2 MSW routes for energy recovery

B1 DIRECT COMBUSTION

Early schemes dating back to the beginning of the century aimed at the recovery of heat energy by burning processed or unprocessed waste by installing boilers in the flues of refractory incinerators have not proven to be particularly successful. Only a small proportion of heat is recovered in this way and there are difficulties in sustaining even-burning which needs to be overcome by drying the waste or adding quantities of fuel like coal or heavy oil. A much improved version of this approach is the construction of special water-wall incinerators where water pipes are installed in the walls and floor of the incinerators. More advanced incinerators have moving grates and fluidised bed versions are under development.

Some water-wall incinerators have been in operation in Europe since the 1950s and show distinct advantages. For instance they operate at higher burning temperatures as there is less excess of air, and combustion is more complete. They are less polluting as the heat extraction by the water in the pipes means lower flue gas temperatures. The problem of effluent is much less because of this as the gas is easier to clean by 'scrubbing' by passing through a cold water spray, or by electrostatic precipitation of particulates.

Usually, because of the higher burning temperatures, unprocessed waste is used in its entirety and secondary recovery of metal and glass materials is carried out on the ash, the residue of which is disposed of as a more-easily handled landfill and which may even have a commercial value as distinct from being a liability.

B2 REFUSE-DERIVED SOLID FUEL — RDSF*

The simplest approach to producing RDSF is to sort, dry and screen MSW, that is to recover desirable recyclable materials prior to burning, and obtain better burning from the organically enriched material by reducing the size variation and reducing the moisture contact typically to 16 to 20 percent. Shredding produces even better burning and more easily handled material and this and further processing, including chemical treatments can be used to produce proprietary, standardised fuels.

Methods of sorting include air and water classification — or separation of denser fractions by weight difference while moving through air or water — and ferrous metal separation by using magnets. In the case of water classification wet pulping may be used to obtain more homogenous material and in this case the RDSF would be used for pyrolytic distillation rather than for direct combustion.

MSW processing is, in particular, geared to materials recovery but shredded material is also more easily handled and fed into incinerators. The higher cost of this can be justified by the better quality of the fuel. A typical scheme as used in Milwaukee handles an annual throughput of up to 400 000 t/yr with a maximum capacity of 1600 t/day. The steps following collection are:

Removal of bundled paper and corrugated material
Shredding in a hammer mill

---

*Alternatively: Waste-Derived Fuel — WDF

Air classification into light and heavy fractions
Secondary shredding to 6–25 mm size
Compaction and transport to a local power company furnace

The heavy fraction separated by the air classification is sorted for recyclables — metal, wood, glass, plastic and ceramic, and the remainder used for land-fill. This process converts about 50 percent of the original MSW to RDSF.

Table III compares the composition of RDSF fuels with bituminous coal. In this table 'Ecofuel' is a further-processed dried, powdered and briquetted version and 'Hydrasposal' a pulped version with 50 percent water content.

Table III Composition and heating values of RDSF compared to coal. (After Gordian Associates, 1977)

Percent	Typical RDSF	Ecofuel	Hydrasposal	Bituminous coal
Carbon	28.35	39.7	23.25	69.70
Hydrogen	3.99	5.3	3.2	4.65
Nitrogen	0.53	0.9	0.35	1.40
Oxygen	20.79	32.3	17.15	6.25
Sulphur	0.17	0.15	0.1	4.00
Ash	16.17	11.5	5.65	11.50
Moisture	30.00	10.0	50	2.50
Other (chlorine)	–	0.15	0.6	–
Total	*100*	*100*	*100*	*100*
Gross heating				
Value (Btu/lb)	4620	6900	4200	11500
(MJ/kg)	9.7	16.0	9.7	26.6

Though MSW/RDSF is clearly a useable resource for energy generation through steam or hot water, its use depends very much on the local situation and in most cases is still only economically marginal in terms of the investment necessary. There are doubts about the costs and reliability of the systems available and the low temperatures and pressures achievable even by water-wall incinerators severely reduce the utilisation value of the steam generated.

Doubt about costs of operation is likely to be at least partly relieved as prime fuel supplies become more expensive. However, to some extent this will be offset by increased capital costs of plant which themselves are related to the basic energy costs. A key factor in the economics of operation is the cost of the transport of the MSW fuel and most viable schemes require the use of the derived energy at, or close to the site where it is generated. Ideally waste disposal and power generation facilities should be combined together at one site, but for reasons of historic organisation, this is not usually so.

SEWAGE AND MUNICIPAL WASTES

Steam generated by MSW combustion is usually at low or medium pressures — typically at 400–625 psig (28–44 bar), but can go up to 900 psig, (63 bar). This is due to basic design and problems of corrosion and it limits the value of the energy resource. At higher temperature some can be used to generate electricity, but at low temperatures and pressures it is only useable for space heating, process heat in manufacturing and driving machinery. In these cases the energy can only be used close to the site where it is generated as neither steam or hot water travel well. There are particular problems of matching supply to demand. Waste plants need to be in continuous operation as waste is generated constantly. Steam and hot water may be required only intermittently by industry unless it is working round the clock. Space heating requirements are seasonal. Continuous 'base-load' customers are thus to be preferred, however to be fully reliable to supply this MSW schemes may require back-up fossil fuel boilers, thus increasing the capital cost. Steam mains are also a highly expensive item to install.

Hot water and steam from incinerators is used for district heating schemes in Doncaster and Newcastle in the United Kingdom, but as noted earlier, schemes like this tend to defray the high costs of waste disposal by incineration rather than being profitable in their own right.

B3 DIGESTION

Digestion of MSW to produce a methane-rich gas is essentially the same process as anaerobic digestion of sewage sludge or waste green plant matter. Though the process is the same, the predominantly cellulosic material itself is different so the technology used for municipal waste water is not appropriate. In particular cellulosic fibres tend to form a scum which requires breaking up by mechanical method or gas agitation.

Another difference affecting the technology is the huge quantities of material that may be involved, requiring large-scale plant. A method of recovering gas which avoids this requirement involves the tapping of accumulations of MSW deposited as land-fill. This is now being practised at a number of cities and is dealt with separately in the following section.

When using digester tanks some preprocessing of MSW is necessary, using the same methods as for RDSF processing. However, the recovery of non-digestibles is prior to the energy extraction rather than from the residue as may be the case with combustion, and fine grinding to produce a 'fluff' RDSF is an advantage. Also, water classification methods can be used because digester feed material is aqueous.

A typical system to produce digester feed described by Wright et al. (1978) which processes MSW through the following modular stages:

Size reduction through a shredder
Ferrous recovery by a magnet

Air classification to separate remaining heavy products (mostly aluminium and glass)
Further size reduction through screens
Crushing and eddy current separation
Crushing and flotation separation of remaining high specific gravity constituents

A proportion of sewage sludge is added to provide a nitrogen source for the digestion process

In this case the gas generated from the digester fraction is used without upgrading as an on-site boiler fuel to produce electricity. After digestion the remaining stablised sludge can be used as a fertiliser.

B4 LAND-FILL GAS

A method of disposal for municipal solid waste that is cheaper than incineration, if sites are available and not too far away, is land-fill. Huge tips and infilled quarries or gravel pits are a feature of the outskirts of many cities. In the early 1970s in the United States it began to be realised that the biomass content of these tips represented a significant fuel source, not to be mined, transported and processed elsewhere but, conveniently, to be exploited in place. This is because, under the right circumstances, the land-fill generates methane-rich gas on its own.

Particularly in tips in California where urban, industrial and other developments like golf courses were being established on land reclaimed by or from waste tips, the generation of gas by the underlying material was in a number of cases becoming an embarrassment and even a hazard due to the possibility of explosions. Accordingly, at a number of sites gas collection systems had to be installed and the gas vented or flared off. It was soon realised that the collected gas could be a considerable energy resource and schemes were put in hand to exploit it. An added impetus was the growing shortage of natural gas in the United States.

Tip material can consist of household, construction or industrial wastes. Most of the latter two are not decomposable and may often contain toxins that will prevent the formation of biogas. Household waste on the other hand may be, as we have seen, widely varied in composition with different proportions of kitchen waste, paper and wood.

However, most of the material is cellulosic and the process of biogas formation is similar to that of the digestion of sewage sludge. That is, it follows the route of organic matter being broken down by acid-forming bacteria to give organic acids. These are then broken down in turn by methane-forming bacteria which give rise to a gas containing mostly methane and carbon dioxide.

The composition of the gas as it is generated naturally varies from site to site but typically it contains 40 to 60 percent of methane. The balance, though mostly carbon dioxide, also will contain small amounts of nitrogen, hydrogen sulphide and other gases. The biogas has a medium heating value — typically 400 to 450 Btu/scf (14.9–16.7 MJ/m^3) and is associated with varying amounts of water vapour. The ideal conditions for the generation of this gas as determined by James (1978) are:

    Temperature 29–37 °C
    No air present
    pH 6.8–7.2
    Moisture content greater than 40 percent
    No toxins present

It should be noted that though a high moisture content is necessary, total water-logging is decidedly inhibiting so that flooded sites or tips extending below the water table are not suitable. Ingress of air stops the gas generation process and the requisite anaerobic conditions are more likely to arise in deeper deposits of waste. Indications are that a 10 metre depth at least is necessary for satisfactory sites and preferably 30 metres or more. So far recovery of gas from sites of less than 10 hectares in extent have not been considered in the United States.

Ideally, to prevent gas leakage, tips should be surrounded by impermeable material — for instance in a clay pit — and should be covered with a layer of impermeable top soil — again ideally, clay. In sites on which work has been done it appears that this need not be more than a metre in thickness. In future, if new tips are to be exploited for gas, they can be specially sited and managed for the purpose. The actual tipping can be controlled to exclude non-degradable material and inhibiting toxins. Also, sewage sludge can be added to improve the digestion. Pilot schemes now in operation on deep land-fill sites of 16 to 65 ha in extent yield up to 150 000 m^3 per day of biogas (1 million scf/day) which is equivalent to about half this quantity of pipeline quality gas. Expansion of up to 728 000 m^3/day (5 million scf/day) are planned with lifetimes for the schemes of 8 to 15 years.

The methods of gas collection involve the sinking of sealed wells into the land-fill, connecting these by a lateral collection system and pumping out the gas. In the original venting systems the gas was simply allowed to blow out under its own pressure but where the gas is required to be used it has been necessary to install compressors to maintain a high enough flow rate. Wells, varying in depth from 10 to 40 metres are typically gravel packed and sealed at the top with clay or cement.

In some cases the low Btu gas is burnt directly at the site in specially adapted boilers to provide heating. In others, where the gas is sold off to public utilities and has to be fed into the main pipelines as a supplementary supply, it must be up-graded to the equivalent Btu value of the natural or synthetic pipeline gas

(LNG or SNG). The gas treatment essentially consists of the removal of the carbon dioxide by passing the raw gas through absorption towers. A typical methane gas purification process employs activated carbon and molecular sieves of silicon gel. The CO_2 absorbed by these filters is periodically driven off by heating (thermal cycling) and the whole system, including the gas extraction compressors can be fueled by the gas itself. The main problems that have emerged are from water corrosion in various parts of the collection and processing systems arising from the water content in the raw gas.

The most advanced and successful developments are underway in California where large accumulations of waste associated with large carbon complexes occur in a warm climate. Here the peculiar circumstances of the need to vent gas when urban pressures led to the reclamation of waste tips combined with a natural gas shortage in the 1970s, the two factors combining to make developments appear to be commercially viable.

At the City of Industry site in Los Angeles (Stearns et al., 1978) 30 shallow wells on a 65 ha site developed as a recreation (golf course) and industrial area are expected to provide gas for over 10 years. The gas is used without up-grading as a reserve supply to boilers on site and is estimated to save some tens of thousands of dollars in the purchase of pipeline gas. However, need for the collection of gas to protect the site provided by a large element of subsidy to the project.

The first site, however, to be developed was at Palos Verdes (Los Angeles area) where 8 wells were installed on a 16 ha site in 1970 to vent off gas, and the system converted to sell off the gas in 1975. The scheme, which produces 150 000 m^3/day of gas has a process plant to bring the gas up to pipeline quality — 99 percent of methane — and sell it to the distributing authority.

The Municipality of Los Angeles is taking an active part in these developments and there are a number of other schemes in that area, at other sites in California — a particularly large one at Mountain View, and also associated with other cities in the United States like New York, Staten Island. As well as for pipeline supply and use on-site the gas may be used in local generating stations or sold to chemical refineries in different cases. In the United Kingdom work by private companies and local authorities is underway using land-fill gas in brick kilns.

Unlike many biomass schemes these appear to be sound propositions in a number of cases, particularly as they have an essentially free feedstock and other benefits like increased land value are associated with them. Clearly, there is considerable scope for the development of other waste-tip sites of sufficient size throughout the world. However, the schemes are by nature of large size — the Sheldon–Arleta Gas Recovery Project (Sun Valley, Los Angeles) for instance cost in the order of $2.5 million for 14 wells (Bio-Energy Directory, 1978).

B5 PYROLYSIS AND GASIFICATION OF MSW

There are a number of physical-chemical methods of heating MSW in the absence of or with a modified supply of air to give rise to gas or liquid fuels. Usually there is a gas stage on the way to a more energy-dense light alcohol or pyrolysis oil fuel. Though this pyrolysis or hydrogenation gas usually has a higher calorific value or energy content than digester gas, the energy gain in the product is always off-set to some extent by the energy consumption in providing the heat for the process.

The principles of gasification are the same as for coal gas generation processes for which have been in operation since the early nineteenth century. Any carbonaceous material can be treated at high temperatures to react with air, oxygen or steam to give a gas product composed mainly, and in varying proportions of methane, carbon dioxide and hydrogen. Some methods have the advantage of being autothermal in that they use their own heat of combustion to provide the energy for the process.

The disadvantage of MSW, and other biofuels, for these processes is that these have a low calorific value in the first instance and tend to produce smoke and tarry residues. An advantage is that usually, the material does not contain significant quantities of sulphur, though the chlorine content — coming to some extent from plastics — may be a problem. Because of this it may be necessary to add lime to the raw material to neutralise it.

Two autothermal gasification routes are particularly suitable for MSW. These are:

(*i*) Gasification with air and steam to provide partial oxidation giving low Btu gas: 100–300 Btu/scf (or 3.7–11.1 MJ/m^3).

(*ii*) Another partial oxidation reaction with oxygen and steam to give a medium Btu gas of 300–400 Btu/scf (or 11–15 MJ/m^3).

Air-free gasification with hydrogen to give pipeline quality gas of ±1000 Btu/scf (or ±37 MJ/m^3) is considered less promising, but this highly reducing route is used for pyrolysis oil production.

Co-current moving bed gasifiers are being developed by Twente University in the Netherlands to be fueled by all kinds of solid waste including MSW. The process involves injecting air into the lower part of a top-fed reactor causing a circulation current above the inlet in which drying and breakdown of the fuel occurs (the moving bed), with pyrolysis and partial oxidation. The finely divided material then falls to a reducing zone from which the hot gas is extracted.

Other designs being developed at other centres use fluidised or entrained suspension of the material by injection of air or oxygen into the base of the reactors. Though these are essentially small-scale systems it is projected that reactors of the Twente type could handle up to 100 t/day. This is still not large-scale waste treatment but the gas outputs would be suitable for both heating and power generation.

A large-scale system being operated at Baltimore, USA, used 600 t/day shredded fuel input and is of the first type described above, using about 40 percent of the air that would be needed for complete combustion. The gas produced has 120 Btu/scf (4.4 MJ/m^3) and is sold to a local gas and electric utility company for direct burning to raise steam for district heating.

Of the second type, the PUROX process operated again in the USA in plants of up to 2000 t/day capacity uses a pure oxygen rather than air feed. The method is similar to that of the Twente University process with coarse shredded MSW being fed into a vertical shaft reactor to fall through a rising current of hot gas. The gas from this system is rated at 350 Btu/scf (13 MJ/m^3) and at different plants is used either as a fuel gas direct or up-graded in a methanisation plant to pipeline quality at between 900 and 1000 Btu/scf (33–37 MJ/m^3). It cannot be used without processing as domestic gas as it has a high carbon monoxide content. Alternatively, it can be used as a synthesis gas for methanol or other high energy fuels or chemical feedstock.

A number of experimental processes take fuel synthesis one stage further. The production of pyrolysis oil usually needs high temperature processes with a hydrogen atmosphere. The output however can be high quality light oils with low suphur contents.

B6 ECONOMIC CONSIDERATIONS

The progression of options for the reclamation of energy from MSW — from direct combustion to digestions to gasification and production of liquid fuels — produces increasingly energy-dense results. However, the technical solutions become progressively more complex and expensive and at the high Btu gas and liquid fuel ends high energy inputs are required. Though in some cases most of this can be provided directly from a proportion of the MSW feedstock itself the use of oxygen or hydrogen in processes represents a considerable 'outside' input of the energy needed to generate and compress these gases. The factors of energy accounting are therefore directly influenced by basic conventional fuel costs as is the economic viability of energy or fuel production by any scheme. A further factor that must be considered is the nature of recovered energy or fuel insofar as it affects its eventual use. Steam and low temperature and pressure and low Btu gas do not fit readily into energy supply networks and up-grading to more convenient forms is necessary, e.g. to electricity via additional equipment. Alternatively, its most effective use is in localised special purpose applications like district heating. In the first instance an additional energy conversion loss enters into the calculations, in the second an appropriate scheme must be located close by, or alternatively the scheme must be located close to the energy user. As the considerations are, and will remain, secondary to waste disposal, the viability of any scheme depends very much on the local circumstance. Because of this a variety of different schemes may be viable in different situations alternatively depending on capital costs of installation and basic fuel cost savings achievable, which are dependent ultimately on basic energy costs.

B7 HYDROLYSIS AND FERMENTATION

It is certainly conceivable to hydrolyse MSW by a short-time high temperature treatment with mineral acid according to the hydrolysis technology discussed in chapter 2. A separation process is necessary prior to applying this process because hard solid objects in the feed would hamper the operation of the plant and metals would dissolve in the acid at the temperature used. The fraction that is desirable as feedstock is predominantly the cellulose one, especially the paper and board fraction, and textiles made of natural fibre, though wood, garden wastes and food wastes, are acceptable and will yield fermentable sugar. Plastics, leather and rubber, although they come into the organic fraction, are not desirable, even if they are up to a point unavoidable, because they are either inert in the reaction or give rise to unwanted reaction products.

Proteinaceous wastes, such as leather, do tend to have an interfering effect by promoting side reactions between sugars and amino acids, reducing the sugar yield. Ideally, then the initial sorting step would concentrate the cellulosic fraction as much as possible. This would then be slurried with water and mixed with dilute sulphuric acid, the solid/liquid ratio being between 1:10 and 1:4 (Converse et al., 1971; Meller, 1968) using approximately 1% sulphuric acid and a reaction temperature of between 200 and 300 °C applied for less than one minute. When this is applied to MSW after removal of say, 25% of metal, glass and other inorganics, the feedstock, representing some 75% of the original waste can be expected to yield approximately 30% of the MSW dry weight as sugar, or 40% of the separated organic fraction. On this basis the sugar that could be produced by this means in the European Communities, for example, from the annual 170 Mt/yr of MSW (probably about 100 Mt of dry matter) would be 30 Mt/yr, or almost three times the annual output of the indigenous sugar industry. However, the sugar would be glucose, not sucrose, and would need to be subjected to an isomerisation step, to convert it to a glucose/fructose mixture, before it could be used as a sweetening agent. Other uses would be as feedstock to produce animal feed (single cell protein) by fermentation or chemicals by fermentation and/or chemical transformation as already described for the case of straw hydrolysis in chapter 3. Otherwise, the sugar could be anaerobically digested to yield methane or anaerobically fermented with yeast to yield ethanol as liquid fuel.

The economics and energy balance considerations involved in the last two options are also considered in chapter 2.

[9]
Aquaculture and marine harvesting

Aquatic plant species exhibit some of the highest recorded growth rates and overall productivity levels and therefore they have attracted attention for biomass production for energy use. Prospects considered vary from the management and harvesting of wild stocks to sophisticated schemes for farming freshwater and marine macrophytes and microalgae.

The high productivities mainly result from the less variable aquatic environment and the fact that the plants are for the most part totally immersed in nutrient-carrying medium. However, overall growth rates are just as much limited by total solar energy received as they are for terrestrial plants.

A Freshwater macrophytes

The plants considered under this heading are fully aquatic water weeds and rushes having much of their biomass above water level. They do not include wetland tree species like willows and poplars. As indicated in chapter 1 communities of higher plants growing in a freshwater aquatic environment can have particularly high yields. For instance Westlake (1963) suggests that northern European aquatic environments can yield over 30 t/ha/yr. In warmer climates higher yields are likely to be achieved more often.

Clear manifestations of high biomass yield can be seen in the huge reed swamps of the Sudd in Southern Sudan (main species *Cyperus papyrus*) and the Pharagmites/Typha beds in Southern Iraq. The Sudd has over 1 million ha of papyrus swamps and it has been estimated that this represents up to 150 Mt of plant material. Water hyacinth (*Eichhornia crassipes*) by its prolific growth has caused considerable problems in freshwater. Its introduction by accident into Africa has caused major blockage of waterways, particularly in the Congo and the Nile. In the Far East, however, and other places where there is a high farming population density it is more easily controlled and it has some uses as a crop mulch, low grade fertiliser and animal feed. Reported yields for some aquatic species are given in Table I.

Because freshwater is much more limited in area than agricultural land or salt water, opportunities for the collection or deliberate cultivation of water weeds

Table I Annual growth rates of freshwater macrophytes (dt/ha/yr)

Berula/Ranunculus	United Kingdom	8.5*
Ceratophyllum demersum	Sweden	9.0*
Phragmites	United Kingdom	7.5–13.0†
Typha	United Kingdom	10.7†
Typha	Minnesota	25*
Sagitaria	Florida	27
Eichhornia crassipes	Louisiana	15–44*

*Quoted by Westlake (1963)
†quoted by Newbold (1971)

are quite limited. Harvesting of areas large enough to give useful yields will also require special, mainly new techniques. A limited amount of development in this field has taken place through the development of some reed bed areas to supply paper mills — notably in Romania and Iraq. Most equipment developed for cutting water weed however, is used to remove nuisance weed rather than to harvest it as an exploitable resource. At the smaller scale a variety of possibilities exist in the processing of waste water, with the production of biomass being an add-on benefit to its main role in reclaiming nutrients from municipal, industrial and agricultural wastes, and 'polishing' water for re-use. Water hyacinth figures largely in such schemes in the southern United States where weed ponds are incorporated into a number of municipal waste treatment plants and are the subject of long-term study. They are also under evaluation by the National Aeronautical and Space Administration (NASA) at their research station in Mississipi with a long-term view for their eventual use in hydro-organic recycling technology for use in space stations. Experiments have been going on since 1975 and one of the promising features of this research is the ability of the hyacinth to absorb heavy metals from industrial waste. Growth rates in warm enriched sewage water can be very spectacular and wet harvested yields very high, particularly as the wet material contains about 93 percent water.

Peak growth rates from the NASA Bay St Louis experiments have been reported as up to 20 per wet tonnes per hectare per day (Bio-Energy Directory, 1979). Even allowing for the drained-off weight being half the wet-harvest weight and for only 7 percent of this being actual plant material and the rest water, the daily growth rate is still an extraordinarily high 0.7 dt/ha/day. This must be an optimum daily figure that certainly cannot be extrapolated to an annual yield which would be over 250 dt/ha/yr. Results from the Woods Hole Experimental station in Massachusetts indicate a dry matter yield of 25 t/ha/yr is more likely. In field conditions in Singapore where hyacinth is grown in almost ideal year-round climatic conditions in fish ponds used to recycle waste from intensively-reared pigs and used as a low grade pig food, yields of only about 11 t/ha/yr are reported (Chin and Goh, 1978).

Other species being investigated in this context in the United States are *Lemna minor* (duckweed) and *hydrilla verticilliala*. These have the advantage of more active growth during the winter months than *Eichhornia crassipes*.

Work on the subject is being carried out in Europe by the French National Agricultural Institute centered in Narbonne in southern France. Here the approach is to use eutrophic urban and industrial waste waters, particularly that arising from the wine and other agricultural industries. A variety of fully aquatic and reed species have been considered as well as algae which are constituents of the totally harvestable biomass occurring naturally in the waste disposal and treatment ponds. Work is being done on growth cycles and cultivation techniques with a view to the development of specifically constructed lagoons of several hectares in extent at existing waste treatment plants in Languedoc, southern France.

The ideal, and possibly only practicable way of obtaining useable energy from freshwater aquatic biomass, is by anaerobic digestion where the high initial water contents are tolerated. No particularly unique features of digester technology are likely to be involved, except possibly in the area of pre-treatment and dewatering. One kilogram of dry hyacinth can possibly yield 15 scf (0.42 m^3) and 1 tonne (424 m^3) of biogas with a 65 percent methane content and with a heating value of 22 MJ/m^3 (595 Btu/scf).

At the rates of growth observed in Singapore it has been calculated that the hyacinth harvested from a 420 ha reservoir could be used to generate 8000 m^3 of biogas per day (Chin and Goh, 1978).

In practice, in some circumstances the costly problem of harvesting can be subsidised, as weed clearance can be a necessary and major item in waterway management, and gas production could be an 'add-on' benefit to water purification. In the southern United States more than $10 million is spent a year on weed clearance, of which *Eichhornia* is the chief element, and benefits arising from the clearance are worth many times this. In the Sudan, where there is a strictly enforced policy to prevent *Eichhornia* floating down the White Nile past the Jebel Aulia Dam above Khartoum, the large quantities of weed that are mechanically harvested are being considered for biogas production.

Unlike forest biomass or seaweed, freshwater aquatics have few possibilities for alternative exploitation. The exceptions are the reeds and sedges which have a potential as a source of paper pulp (papyrus, after all, was the precursor of modern paper). Apart from this there seem to be only low value uses. *Eichhornia*, for instance, can be used as a low-grade cattle feed and mulch or soil conditioner. Such uses are more likely to be significant in the developing countries.

B Seaweed

Seaweeds are the marine macrophytes but with one exception, the eelgrass *Zostera*, they are all algae — simple forms of non-vascular plants. Though they are broadly divided into three main groups; the chlorophytes (green algae), rhodophytes (red algae) and the phaeophytes (brown algae) these essentially colour differences are in no way comparable to the huge diversity in structures found in terrestrial vascular plants.

Like all plants, with the exception of parasites and some saprophytes like the fungi, they require light to grow and because with few exceptions, they need to be anchored on the seabed to grow, they are largely limited to shallow waters, the actual depths to which they are found depending on the clarity of the water. There are few exceptions to this, the only free-floating marine types being *Sargassum* species in the Atlantic, *Phyllopharae* in the Black Sea and a few Pacific species. The type of seabed is also critical to their occurrence as they do not root into soft sand or mud but require hard rock on which to establish their holdfast mechanisms,. At suitable locations seaweeds are found from the inter-tidal zone to the limit of light penentration. In clear waters in Northwest Europe this is down to a maximum depth of 20 metres below the lowest spring tides.

Marine algae occur in a variety of sizes and forms, from filamentous mats to dense clumps and long fronds. In size they vary from microscopic unicellular types to 15 metre fronds in the case of the largest type, *Macrocystis pyrifera*, the giant kelp of the Northwest coast of America. In northern seas it is the large brown types that predominate; the Fucoids (wracks) and *Laminaria* species (oar weed) and it is these and species like *Macrocystis* which show the spectacular accumulations that appear to offer big prospects of energy from biomass.

Though in places huge masses of weed may be washed up by storms and very large beds of very fast-growing weed occur, they are relatively little exploited, being difficult to harvest and handle and being useful for only limited number of purposes. As such, seaweed is thought of by some as being particularly promising as a source of fuel energy as systems to exploit them will be dealing with new types of plant growing in a relatively unused environment and not competing with traditional crops.

Growth rates of seaweed can be very spectacular. Giant kelp fronds off California are reported as growing at a rate and up to 4.5 m per week (65 cm/day), and in terms of bulk rate increases of as much as 20 percent per day under favourable conditions are reported. These are of course maximum growth rates and they cannot be extrapolated into annual yields. The same rules of plant growth apply to marine algae as they do to any other plants, being limited by temperature, light and the availability of nutrients. Light is not only limited by water depth but by shading of adjacent plants, so that growth rates of new colonies will tend to be inversely proportional to the stocking density. Examples of growth rates of some species are given in chapter 1. It should be remembered however that some figures quoted may be calculated from extrapolations of the optimum yield and figures for actual harvested yields may not take account of the fact that the drained off, not even air-dry weight may be only half that of the wet-harvested weight and that the material tends to contain much more water than terrestrial biomass, typically 90 percent compared to 50 percent.

However, the materials may contain high concentrations of carbohydrates and oils which increases their potential energy content. Because of this, however, some material is in demand as sources of valuable chemicals. The lack of lignin in seaweed material also makes it potentially highly digestible or fermentable, though the high chloride content from the seawater poses problems for these processes. Tables II, III and IV, show some of the chemical and energetic characteristics, of giant kelp from California. These are fairly typical for seaweed and it can be seen that the energy content is quite low compared to that for terrestrial

Table II Composition of *Macrocystis* (%). (After Bryce, 1978)

Solids 12.5	(Organic matter) Volatile solids 55–62	Carbohydrate 28	Mannitol	0–14
			Laminarin	0–21
			Fucoidin	0.5–2
		Structure and Pigments 27.5	Cellulose	3–8
			Algin	13–24
			Protein	5–7
			Fat	0–0.5
	Ash 38–45		KCl	28.7
			NaCl	7.5
			Na_2SO_4	4.3
			Trace	4.0
Water 87.5				

Table III Chemical analysis of *Macrocystis pyrifera* (California) (%). (After Chynoweth *et al.*, 1978)

Initial moisture content	88.8 weight
Volatile solids (organic matter)	57.9 dry weight
Residual ash	42.1 dry weight
Analysis of dry materials	
C	27.8
H	3.73
N	1.63
S	1.05
P	0.29
K	14.7
Na	3.5
Heating value (dry weight)	10.7 MJ/kg

Table IV Composition of the organic fraction of kelp (50–60% of dry weight). (After Chynoweth et al., 1978)

	Percent
Protein	29.5
Carbohydrates	
Mannitol	34.5
Algin	26.1
Cellulose	8.8
Laminarin	1.3
Fucoidin	0.4

plants. The reason for this is the high non-organic ash content. The organic, or volatile solids content which is convertible is only about 50 to 65 percent of the dry weight compared to 75 to 95 percent for terrestrial plants.

B1 STOCKS

There are undoubtedly very large stocks of unexploited seaweed potentially available. Individual storms may cast up as much as 10 000 tonnes of weed at a time and from suitable indented and rocky coastlines in western Scotland it has been estimated that over 2.5 tonnes per month could be harvested from each kilometre. Offshore beds of wracks (*Fucus* and *Ascophyllum* spp.) have growing densities of between 50 and 90 t/ha, of which perhaps one third could be harvested annually to allow full regrowth (Chapman, 1970). Offshore stocks of *Laminaria* alone off Scotland are estimated to be over 10 million tonnes. Table V gives estimates of potentially harvestable stocks, worldwide, based on harvested yields. These show over 17 million tonnes. It should be noted however, that all the over 2 million tonnes actually harvested annually in the 1970s was assigned to various non-energy uses.

The main problem for any bulk user of wild stocks is in harvesting. Though boats are efficient in some cases the harvesting from rocky shores where some of the best stocks are found defies mechanisation. There are however a number of industries based on seaweed. The largest and most highly organised of these is on the Californian coast where the kelp beds have been exploited since the beginning of the century and controlled cutting has been carried out by special vessels since the 1920s. Harvested yields here are at a rate of about 2 dt/ha/yr of an average total productivity of about 20 dt/ha/yr (Clendening, 1917): *Furcellaria* is trawled off Denmark at a rate of about 100 t/day.

B2 MARICULTURE

The alternative to gathering wild stocks is deliberate cultivation. This is already carried out in a number of different ways for a number of different purposes, but none involving fuel as an objective.

Table V Potential world seaweed resources (10^3 tonnes). (After Naylor, 1976 and Michanek, 1975)

Area	Red algae 1971–73 harvests	Red algae Potential output	Brown algae 1971–73 harvests	Brown algae Potential output
Arctic	—	—	—	—
Northwest Atlantic	35	100	6	500
Northeast Atlantic	72	150	223	2000
West Central Atlantic	...	10	1	1000
East Central Atlantic	10	50	1	150
Mediterranean and Black Sea	50	1000	1	50
Southwest Atlantic	23	100	75	2000
Southeast Atlantic	7	100	13	100
West Indian Ocean	4	120	5	150
East Indian Ocean	3	100	10	500
Northwest Pacific	545	650	825	1500
Northeast Pacific	—	10	—	1500
West Central Pacific	20	100	1	50
East Central Pacific	7	50	153	3500
Southwest Pacific	1	20	1	100
Southeast Pacific	30	100	1	1500
Antarctic	—	—	—	—
Total	807	2600	1315	14600

In the Far East some edible seaweeds are cultivated as relatively high priced delicacies and for medicinal uses. In Japan and China these include the red 'Nori' and some small kelps. The method of growing is on ropes and rocks which are artificially seeded and fertilised. Yields of these are low, generally not more than 1 dt/ha/yr. Higher yields are obtained from *Euchina* in the Philippines — 13 dt/ha/yr and *Gracilaria* (sea lettuce) in Taiwan — 9 to 10 dt/ha/yr, with possible yields of up to 30 dt/ha/yr being projected (Ryther et al., 1977).

Trials at Woods Hole Marine Research Institute in Massachusetts have involved growing seaweed as food for bivalves, fertilising the growing troughs with sewage. Similar experiments in the Gulf of Mexico use nutrient-rich subsurface waters as the fertilising medium, pumped up to form an artificial up-welling situation. Both large weed and microalgae production have been explored at this facility. (Roels et al., 1976). Similar tests have been carried out in Florida. (Ryther et al., 1977). It is trials of this kind that have given the high yields of up to 30.7 g (dry)/m^2/day under continuous harvesting that have indicated potential annual harvested production rates of over 100 dt/ha/yr, referred to in chapter 1. The technique of growing in troughs on land is practiced commercially with the red kelp *Chondrus crispus* in Nova Scotia where it is grown for its chemical (carrageenan) content.

Off California a more ambitious technique of open sea farming is being attempted with *Macrocystis* with the specific aim to produce the material as an energy feedstock to generate methane. The intention is to establish the weed on floats anchored in midwater and to harvest by cutting in the same way that the natural giant kelp beds are harvested. This is similar in approach to the short rotation forestry coppicing approach on land.

In the natural kelp beds productivity has been shown to be limited by the extreme nutrient depletion in the upper water layers, and the intention is to fertilise the artificial beds by pumping up deep nutrient-rich waters from as much as 500 m. The pumping units and networks of floats are conceived as separate modules. The idea is illustrated in Figs 1 and 2. At the time of writing

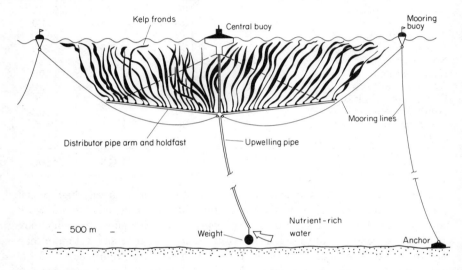

Fig. 1 Kelp farm module. (After Bryce, 1978; North, 1977)

a 1000 square metre unit has been established, as one stage in a development plan that envisages a 200 to 400 ha farm for full development of the process between 1982 and 1987, and a commercial scale 40 000 ha farm by 1992 (Bryce, 1978). Harvested production from these farms is envisaged to be as high as 120 dt/ha/yr. However, 50 dt/ha/yr may be a more realistic target (North, 1977).

The above represents the most ambitious of the mariculture schemes but a considerable amount of other work is being undertaken. This includes studies of growth and nutrient requirements, chemical analysis and fuel generation, and selection of high yielding strains of selected species. This work is aimed at both seaweed for its fuel and other chemical feedstock potential and is particularly

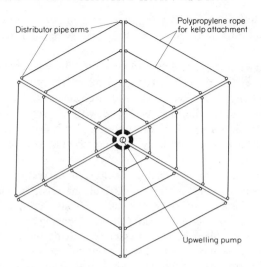

Fig. 2 Plan of the sub-sea portion of a kelp farm module. (After Bryce, 1978; North, 1977)

centred upon *Macrocystis pyrifera*, *Gracilaria folifera*, *Chondrus crispus* and another carrageenan-rich species, *Irideae cordalia*.

The idea of seaweed farms in the open ocean or any management of wild stocks by anything more ambitious than controlled cutting faces the problem of the marine environment which is very severe where most of the main high density stocks occur in high latitudes. In many cases it would entail the development of completely new technologies. Such approaches might be more feasible in calmer waters — the Baltic, Mediterranean and the Adriatic. Though the present relatively poor stocks of weed in the Mediterranean are probably due to nutrient deficiency the Californian work indicates that even there in the open Pacific additional nutrients are essential if adequate yields are to be sustained. (Jackson, 1977). The possibility of introducing faster growing species to these areas might also be considered. However, the proposal to introduce *Macrocystis* to Brittany for the benefit of the algal chemical industry met strong objections from environmentalists and the local fishing industries.

The problems of mariculture on any scale are primarily concerned with engineering costs and growth. Mariculture on land will involve considerable expenditure on artificial environments of troughs, tanks or modified natural lagoons and the necessary pumping and materials handling equipment. In the marine environment the practical implications of large-scale engineering of the type envisaged are almost unknown, though the Californian QAM (Quarter Acre Module) test facility will give some of the answers. The cost of overcoming the engineering outlay can only be borne by exceptionally high yields of seaweed, so essentially, as with any form of farming, the question comes back to harvested yields. Overall mariculture is probably best equated with intensive farming on land in terms of economics.

Like all plants seaweeds require light and nutrients, these last being carried by the water in which they are immersed. The implication from this is that relatively high flow rates are demanded — which is another engineering and cost problem. However, the problem of shading can be coped with by frequent cutting. The high energy inputs in all the activities can only be justified by the sort of high sustained yields that have probably not yet been proved as generally possible on a large scale.

Other problems include the practical difficulties of harvesting, handling, dewatering and processing of what is a rather difficult material, originally containing up to 95 percent of useless water. On the cultural front there are potentionally serious problems of crop reduction by infestation of grazing animals — molluscs and others — and the problem of crop hygiene involving the exclusion of undesirable species. Being, in most propositions monoculture, the plantations will be vulnerable to disease. *Zostera marina* beds in the Atlantic were practically wiped out in the 1930s by a single pathogen species.

B3 FUEL GENERATION

The main interest in fuel generation is in methanation. Work on this (Chynoweth *et al.*, 1978) indicates that yields of 0.4 m^3/kg to 0.56 m^3/kg are theoretically possible from *Macrocystis*, with over 80 percent conversion of the organic matter content. In practical large-scale anaerobic digestion, however, yields are more likely to be of the order of 0.28 m^3/kg. These are good yields compared to other species tested and they compare with other green vegetable feedstocks. The advantage of kelp species is that most of their organic fraction is in algin and cellulose which is almost completely biodegradable by anaerobic digestion.

There may be particular problems of digestion with digestion of seaweed because of the high chloride content in the water associated with it. For this reason it is possible that thermal conversion may offer on alternative route, particularly towards liquid fuel generation with species with high oil contents.

B4 ALTERNATIVE USES

Even if seaweed could be produced in larger quantities it may not necessarily become available as a fuel feedstock as there are a number of competitive users which could have a better claim. The possibilities and their economics are not, however, clear as yet.

Traditional uses of seaweed are in human and animal feed, and agriculture. Commercial growing of some species for food in the Far East has been referred to earlier and small quantities are still eaten in Europe and from time to time seaweed is used as a component in processed foods. Rather larger quantities are used for supplementary animal food, and as unprocessed fertiliser or soil conditioner. These usages are usually relatively small and local, being confined to coastal areas. Some weed is used in processed proprietary garden soil conditioners.

It is the seaweed chemical industries however that give rise to the most important competitive demand. The old chemical industries which were small and local and obtained soda, potash and iodine by simply burning the weed are now unimportant and have mostly died out. Newer industries (although some date back to the turn of the century) are concerned with the extraction of relatively high value chemicals. These include:

Agar A colloidal gel used primarily for laboratory purposes. The largest producers are Japan, Russia, Spain and Portugal.

Carrageenan A polysaccharide complex with a number of uses in the food industry. Only a few tens of thousands of tonnes of this relatively high value product are produced annually from red kelps (e.g. *Chondrus*).

Furcellaran This is an agar-like colloid, mainly produced in Denmark, at an annual rate of several 100 tonnes. the production of 1 tonne Furcellaran takes about 25 tonnes of weed.

Algins These are used in many food industry and other processes. Main producers are France and the British Isles. The chemical can constitute between 35–47 percent of the total dry weight of the seaweed but normally 5 tonnes of weed are need to provide 1 tonne of product.

The main source of these chemicals are shown in Table VI but in addition there are very large unexploited stocks in other countries.

Seaweeds also contain a large variety of other potentially useful chemicals, on which new industries could be based if the opportunities arise through

Table VI Important seaweeds used for chemical feedstock

Seaweed	Location	Chief chemical extracted
Ascophyllum	France (Brittany), Scotland	Algin
Chondrus	France (Brittan), North America (E. Coast)	
Euchina	Philippines	ι-Carrageenan
Fucaloids	Ireland, Scotland	Algin
Furcellaria	Denmark (Baltic)	Algin, Carrageenan, Furcellaran
Gelidium	Ireland, Japan	Agar
Gigartina	North America	Carrageenan
Gracilaria	Italy (Adriatic)	Agar
Laminaria	France (Brittany), Ireland, Scotland	Algin
Macrocystis	North America (W. Coast)	Algin

Table VII Estimate of world production of seaweed 1973. (After Naylor, 1976)

Location	10^3 tonnes wet weight	Value $ million
Japan	654	563.0
China	700	130.0
Korea (Republic)	224	45.0
USA	126	1.9
Brazil	103	2.0
USSR	100	5.8
Ireland	44	1.3
Norway	75	0.9
France	60	1.2
Spain	47	3.4
Canada	40	2.0
Mexico	37	1.0
UK	24	0.4
Chile	27	1.4
South Africa	24	1.0
Argentina	24	1.2
Portugal	20	1.5
Denmark	11	0.5
Morocco	8	0.4
Others	47	?
Total	*2400*	*765.0*

demonstration of economic harvesting and extraction processes. New processes of destructive distillation (pyrolysis) are able to recover volatile oils, paraffins and naptha, ammonium sulphate and calcium acetate.

The world industry based on chemicals from seaweed is small, but growing. The production and value of seaweed harvested by the industry in 1973 is given in Table VII.

The highest harvested rates appear to be from the California kelp beds where up to 700 wet t/day can be brought in by the specially-built cutting vessels. The main problem for the industry is however, in the economics of collection. The impetus towards more effective methods of production will quite likely come from the chemical industry.

B5 PROSPECTS OVERALL

Exploitation of seaweed for energy appears at first to be promising — some species exhibit very high growth rates and there are large unexploited stocks. Further the sea areas in which they could be cultivated or managed are not under the sort of pressures that land resources are under. However, we have seen that apart from competition for the stock if it becomes economic to collect more of it, the engineering problems and costs of large-scale exploitation are very large indeed. This alone will ensure that any development towards large-scale utilisation, for whatever purpose, will take decades to implement.

C Microalgae

Microalgae grow naturally in freshwater, seawater or brackish water. As for the most part it is not practicable to harvest this material in the wild and the techniques of cultivation are not fundamentally affected by the salt content of the water used, no distinction is made in this section between the types of different origins. The plants themselves may be microscopic or submicroscopic, single cell, filamentors, or colonial types. They are capable of very high growth rates.

We have seen in chapter 1 that the production rates of phytoplankton in open ocean environments cannot be expected to be above 400 g C/m^2/yr, or 5 t/ha/yr biomass and clearly the algal biomass here is too dispersed to be collected directly. There are few natural situations where algae could be considered for collection in quantity, possibly in some lagoonal situations. For instance mats of the colonial algae *Cladophera prolifer* occur off Bermuda and exhibit peak growth rates which could extrapolate to 24 t/ha/yr, though mean production rates are more likely to be in the range of 4–6.5 t/ha/yr (Bach and Josselyns, 1978). Here, growth in sheltered situations is assisted by nutrients from sewage disposal. In freshwater some similar situations arise but 'blooms' of algae in naturally eutrophic ponds and lakes tend to be transitional.

Because of the limited 'natural' opportunities the approach to the exploitation of algal biomass should be by deliberate cultivation on a large scale. Some commercial schemes are established but these are not intended for energy production. On the other hand a variety of approaches to algae cultivation energy production have been proposed or are being experimented with, some for species like *Chlorella* and *Spirulina* are grown for their food and medicinal value. The farms, mostly in North America and the Far East are small, but produce products that are of relatively high commercial value, against which use for fuel generation cannot compete. In consequence total amounts harvested tend to be small, less than 1000 tonnes a year, though high productivity rates are achieved of up to 80 t/ha/yr (presumably wet or drained-off yield).

In the laboratory and in trial ponds maintained at optimum conditions, much higher yields have been obtained. For instance Goldman and Ryther (1977) report several claims for growth rates of up to 30 g/m^2/day dry yield, a yield which if extrapolated would give nearly 110 dt/ha/yr. The authors suggest that these are the maximum yields practically possible, but Soeder (1976) indicates that up to 170 dt/ha/yr might be possible, though this figure is from extrapolations of laboratory growth rates. Others claim growth rates of up to 50 g/m^2/day which would be over 180 dt/ha/yr if extrapolated to continuous growth.

However, these are exceptional figures and even such figures appear to exceed the theoretical maxima, but the production rates may be assisted by the favourable environment, as the algae are bathed in what can be regarded as a nutrient solution. Algae are believed to utilise the C_3 photosynthetic mechanism rather than the more efficient C_4 process of some terrestrial plants. However, overall their typical energy conversion efficiency of 5 percent makes them, as a class of plants, generally more productive than terrestrial types. For laboratory conditions realistic growth rates are probably only 20 to 30 $g/m^2/day$, which in a 300 day growing season would be 60 to 90 dt/ha/yr. For outdoor cultivation this sort of yield could only be obtained in a continuously warm climate and high insolation would be essential. For southern European conditions, Soeder (1976) quotes a possible 45 dt/ha/yr for a 220 day growing season.

The larger scale commercial production facilities probably give a better idea of practically achievable yields. Tsudaka et al. (1978) refer to the 80 t/ha/yr already mentioned for the filamentous types *Chlorella* and *Spirulina*. Algal production schemes for animal feed in Italy have produced up to 40 dt/ha/yr and 50 dt/ha/yr is the assumed production figure for some similar schemes under consideration with up to 70 dt/ha/yr projected (Palz and Chartier, 1980).

The problem of the proportion of water in reported yields is common in the literature. It is particularly important with algae as the wet harvested weight can be twice that of the drained off weight. Air dry material, sometimes reported as 'dry' can include between 4 and 10 percent of water. A common form of reporting laboratory results is as 'freeze dry' weight, which may approach the 'bone dry' used for wood and other terrestrial biomass.

Research into the mass production of algae has been going on since the 1950s but apart from production for the specialist food and medicinal markets, economic systems have not yet been demonstrated. Many methods of culturing and harvesting have been tried, often as part of waste water treatment facilities. In the United States experimental systems at a relatively large scale have been in existence for some time and according to Oswald et al. (1977) reasonably sound cost data is available. The main problem appears to be that there is no proved method of harvesting to give a minimum cost-effective efficiency of 80 percent.

A survey by Tsudaka et al. (1978) indicates that little more than 1000 tonnes of algae (mostly *Chlorella*) was being produced commercially in the Far East in the late 1970s with few individual farms producing more than 100 tonnes a year. Pond or tank sizes in these farms vary from several hundred square metres to 16 000 m^3. These are mostly intensive production facilities involving circulation of the water and the addition of carbon into the system, typically in the form of carbon dioxide gas. Costs are high, algal protein costing up to 10 times more to produce than soya protein food equivalent, but the justification for the production is the high price commanded by the *Chlorella* as a health food.

In Europe a considerable amount of work is being carried out at a number of centres, mainly to explore microalgal production as a food source. Mostly this has been small-scale laboratory demonstrations on species like the unicellular *Scenedesmus* and the filamentous *Spirulina*. As the best conditions for algal growth are in more sunny regions, much of the technology is aimed at developing countries. Very high yields have been obtained and there are prognostications of cost-comparability between algal protein and soya bean protein. The results have lead to a number of propositions for mass production in facilities akin to chemical plant.

One scheme proposed in Italy involves the production of *Spirulina* inside a network of plastic tubes laid out in an arrangement similar to that of a solar collector. This approach avoids some of the high costs of permanent pond construction and being a closed system can avoid pollution and contamination problems, loss of water by evaporation and allows precise control of additives. Small-scale trials of this system indicate that yields of 50/ha/yr or more are possible. The economics suggest that this algal production system could be viable for animal feed but not for energy generation at current fuel prices (Palz and Chartier, 1980).

Another scheme proposed by European groups is specifically for algal biomass production to generate energy by mariculture on land. These are developments intended for southern Europe and adjoining desert countries where the approach does not suffer the disadvantage of requiring large quantities of freshwater and where there are considerable areas of low-value coastal land on which open ponds can be constructed, or better still there are large saline natural lagoons. Experimental ponds have been constructed in southern Italy and a programme of growth, fertilisation and cultural trials are being carried out with an aim to establish a routine production rate at least 50 dt/ha/yr, with a possible target of 70 dt/ha/day. Trials were started with the green unicellular algal *Dunaliella* sp. and the blue-green unicellular type *Aphanothece* sp. These cultures have had a fate common to most of long-term cultures in open environments in that they suffered an almost complete change of composition due to contamination by other species and were also affected by infestation of diatoms which feed on the algae.

This European experience is shared by rather longer standing work with open ponds in Israel (Shelef *et al.*, 1976). Here shallow 'high rate' ponds 35–50 cm deep are used to process municipal waste waters which provide the nutrients for algal growth. Trials have been conducted on pond sizes up to 20 000 m^3, and the harvested and dried algae used for fish and chicken feed. A plan of the type of photosynthetic pond used is shown in Fig. 3. Waste water is circulated continuously along the channels separated by the dividers in the pond and is retained in the system from 2 to 8 days depending on the climatic conditions. The ponds are of low cost construction and the interesting prospect is raised that if, in suitable areas, they were to be constructed of compacted clay, on their eventually becoming silted up they could be readily drained and reclaimed by ploughing

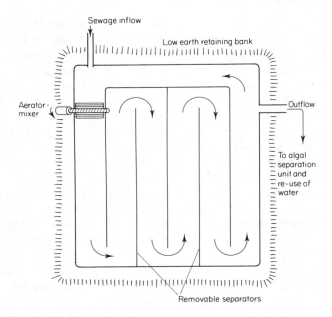

Fig. 3 Plan of a high rate pond suitable for land reclamation. (After Shelef *et al.*, 1976)

as high quality agricultural land. New ponds would then have to be constructed but the disadvantage of their large land requirement would be offset by the increase in value of the reclaimed land. This approach seems particularly attractive in many desert countries where there are many large areas of saline clays which would be leached and reclaimed if fresh waste water rather than salt water were to be used.

The predominant algae identified in the ponds are mostly unicellular green types — *Oocystis solitaria*, *Photocoins* sp., *Scenedesmus dimorphus* and *Micractinium quadrisetem*. Though there is marked seasonal variation in species composition, this does not appear to affect yields markedly harvesting carried out by flocculation of the algae with alum (aluminium sulphate — $Al_2(SO_4)_3$) or ferric chloride (FeCl) at rates of 70 to 120 mg/l. Though this contaminates the biomass, up to 4 percent of alum in the material does not appear to affect its value as animal feed and may not be a major obstacle to fermentation. The dried harvested material contains between 4 to 11 percent of water and yields of up to 110 t/ha/yr have been reported.

Continuing work in the United States, where the original concept of the shallow 'high-rate' pond was established, is concerned particularly with large-scale facilities as part of sewage treatment systems. However, though some algae lagoons have been in operation for a number of years the technique is not sufficiently proved to interest most waste disposal authorities.

Though a number of speculations have been made on the possible contributions of energy from algal biomass, apart from the unresolved technical and cost problems, the fundamental limitation is the provision of nutrients. Added nutrients and carbon in the form of CO_2 or other organic form are essential for adequate levels of production. If sewage is considered as the main source of nutrients, according to Oswald and Beneman (1977), algae fed by the entire sewage output of the United States could only supply about 0.2 percent of the US energy requirements. Added nutrients from other sources will probably be too expensive, but there is the possibility that blue-green algae species which can fix nitrogen from the atmosphere could be used and the CO_2 fraction of the biogas from the digestion of harvested algae.

Oswald et al. (1977) suggest that because of engineering and operating costs the minimum size of a high rate oxidation pond system to produce an economic yield is of the order of 250 ha (1 square mile), with a maximum potential in the United States of 2500 km^2 — determined by land availability and water supplies. Sixty-five km^2 (25 square miles) of ponds would be needed to process the sewage of 1 million people. Very large agricultural benefits are possible by rotating these large pond areas and reclaiming land that will be considerably up-graded by levelling and the incorporation of nutrient-rich sediment. Table VIII shows the range of size and characteristics of a variety of potential schemes.

Though large-scale algal oxidation ponds are in operation for sewage treatment the methods of harvesting which are necessary if the biomass production is to be used cannot be said to be satisfactorily developed. This is the main problem and obstacle to the development of full-scale systems. Though a number of methods have been tried or are in use on the small scale, none have been proved as economic components of large-scale systems. According to Oswald and Beneman (1977) there are as yet no low-cost methods. Centrifuging, used in laboratory work, is too consumptive of energy; chemical sedimentation results in contaminated material which may affect its behaviour in digesters; filtration may proved to be effective but only works with filamentous species.

The consideration of this fact leads onto the next major problem, that of species control. Species composition change over time, partly at least associated with seasonal changes in conditions, appears to be an unavoidable feature of open-air culture where conditions are difficult to control and natural contamination is inevitable. For this reason some of the algae farms in the Far East are operated indoors under closely controlled conditions.

It may be possible to exercise some measure of control through the chemical characteristics of the growing media. For instance some filamentous algae prefer alkaline conditions. However, not enough appears to be known about this topic to be able to safely predict sustained yields from desired species.

Table VIII Suggested algal bioconversion systems in the United States. (After Oswald et al., 1977)

Approx. size range (km^2)	System	Location	Water source	Nutrients N.C.P.	Products/benefits
2.5	Oxidation, ponds	Near urban areas	Liquid wastes	Liquid wastes	Methane, fertiliser, O$_2$
2.5–25	Advanced waste water treatment	Near urban areas	Liquid wastes	CO$_2$, liquid wastes	Methane, fertiliser, nutrient removal
7.5–75	Complete waste water treatment	Near urban areas	Liquid wastes	CO$_2$, liquid wastes, N$_2$	Methane, fertiliser, evaporative waste water disposal
12.5–125	Agricultural fertiliser, energy	Rural areas	Drainage, irrigation	CO$_2$, N$_2$, mineral fertiliser	Methane, N fertiliser, drainage water disposal
25–250	Nutrient integration	Waste lands	Any locally available source	CO$_2$, recycled nutrients	Methane

Sustaining the very high levels of productivity essential to the concepts requires high fertilisation levels. There is a limit to the idea of 'free' fertiliser from sewage. Though each litre of sewage sludge can typically grow 250 g of algae, as Oswald and Beneman (1977) claim that this source could only allow production of 0.2% of the current United States energy requirements, fertilisation in addition to this would be a considerable added cost. Available carbon is the chief limiting nutrient and in a number of proposed schemes CO_2 injection is obligatory. This, of course, means more engineering and running costs.

So far, the only seriously considered energy production route is to methane by digestion, and pyrolysis to produce liquid fuels. Wet harvested material is an ideal digester feed, and the problems of salinity or alkalinity are probably not insurmountable. Most digester experiments have been carried out at laboratory scale only, the possibilities of the larger scale tests being limited by the availability of sufficient quantities of harvested material. The attraction of the alternative possibility of liquid fuel production by pyrolysis arises from the high oil content of some of the microalgal material.

An alternative that is receiving theoretical attention is the production of hydrogen fuel by photolysis by growing algae. This process which is carried out naturally by some species of algae and bacteria splits water into hydrogen and oxygen gas by use of the enzyme hydrogenase. Experimental work (e.g. reported by Beneman, 1978) has shown that cultures of green algae species like *Anabaena cylindrica* produce hydrogen at a rate of 0.4 percent of the solar conversion efficiency, and possibly to as high as 2 or 3 percent. Calculations from this show that it could be possible to produce up to 0.65 litres of hydrogen per m^2 of culture medium surface area in a favourable climatic situation. It could be ten times this if an assumed theoretical maximum solar energy conversion efficiency of 20 percent could be approached.

The advantages of producing hydrogen in this way as a contribution to a possible 'hydrogen economy' are:

The method operates at low temperatures — 10 to 40 °C, compared to chemical methods of hydrogen production which operate at 400 to 1000 °C. It is thus not highly energy consuming.

The necessary inputs, solar energy and water, are cheap.

The process produces a clean burning fuel and produces no pollutants.

On the postulated yields Mitsui and Kamazawa (1977) calculate that a hydrogen producing algal culture 1 metre deep and occupying a 215 km^2 area could supply the total US needs for hydrogen in the year 2000. The culture areas could be in salt water areas, avoiding the problem of competition for land.

As with some of the other energy technologies, practice is well behind the possibilities and large-scale hydrogen production by algae has to be proved to be both technically feasible and economic.

Overall the problems of effecting microalgal energy schemes are rooted in engineering costs. There is little practical engineering experience to date from which to determine the economic feasibility of particular prospects. In many cases the energy required for mixing and pumping to achieve acceptable yields appears to be too high in relation to the output. For these reasons it is possible that the true potential of the concept will have to be worked out where costs are effectively subsidised — in particular in waste treatment schemes. Alternatively microalgal production may be developed by the food industries. In this case, they could be dealing with a higher value product than if it were intended for energy use and only residues would be available for that purpose.

[10]
Overall prospects and limitations

We have now seen a variety of different versions of the energy from biomass possibilities, dealing with potential feedstock sources, technological feasibility and the more difficult question of deciding the economic viability. We have seen the range of possibilities and the individual directions of research that will determine, ultimately, the question. It is now possible to compare these and get some indication of the practical status of realising the fuel and energy potential by the various routes.

A number of other broad questions arise if biomass is to be exploited in quantity in this way. These concern the overall picture of world energy consumption, changing patterns of usage arising from the increase in the cost of 'conventional' fuels and changing technology; doubts about the continuity of supply of the previously dominant fuels, and their environmental effects. Several paths of alternative energy supply are being explored and, in part adopted, of which biomass energy is but one. Though the way things will go will not become apparent for a number of years, major changes in energy supply and use are inevitable. Even the nuclear solution to supply a large part of world electrical energy needs economically, and at a socially acceptable cost, cannot be a complete answer as electricity is not a flexible transportable fuel. The need for gas, liquid and even solid fuels will remain. Because of the diversity of potential solutions, each having merit in a particular context of energy supply or exploitation of a resource the most probable scenario will be one of great diversity including a high degree of local variation in approach, and one in which biomass will play a significant, if not a dominant role.

The question of energy from biomass is also bound up with what is becoming known as the bio-technology revolution, in which biochemical processing will become a powerful factor in chemical and food production. The use of living organisms to synthesise chemicals has many attractions, one of which is their ability to use diffuse, low-temperature energy. As well as competing for food and chemical production many bio-energy prospects are likely to arise in association with the use of this technology. The bio-technology revolution also promises, through genetic engineering, new varieties of plants with higher growth rates, resistance to disease, adaptation to specific and unusual environments, and possibly higher energy densities in their tissues by producing high yields of fatty oils. One aim of particular importance is the implanting of the ability to fix atmospheric nitrogen which can make plants partly self-fertilising. This ability is present at the moment in the Leguminosae among the higher plants and in some blue-green algae.

A Possibilities

Though large-scale energy plantation type schemes of various sorts are envisaged, most immediately practical biomass prospects are relatively small in scale, localised in application, and distinctly 'resource limited'. Most of the immediately practical schemes are concerned with waste material of various sorts. However, the overall opportunity for fuels from biomass has given rise to a number of concepts for world energy solutions based on particular alternative fuel products which are broadly comparable with — and sometimes put up as alternatives to — the 'nuclear solution'. The practicability of the ideas apart, their great attraction is that the basic solar input appears to be free, the fuel feedstocks are non-depleting and renewable, and the solutions appear to be environmentally acceptable — in some cases even desirable. Their non-polluting effects and low adverse environmental impacts are contrasted with strip mining, radiation hazards and the increase of sulphur dioxide and carbon dioxide in the atmosphere arising from fossil fuels.

The complexity of industrial society and the way it has evolved indicates that the unavoidable changes in the energy supply and use patterns coming in the next decades will not be simple and 'one-fuel solutions' will be very unlikely. However, the consideration of these as the ultimate consequences of following individual lines of development are of considerable interest as they emphasise various general objections to biomass schemes as a whole.

A1 THE METHANE ECONOMY

We have seen that at present probably the most successful, and at any rate the easiest, route for fuel production from biomass gives methane-rich biogas via anaerobic digestion. Sewage sludge, municipal solid waste, farmyard manure and green plant matter can all be digested with a minimum of energy input to produce large quantities of gas. Digestion is also the favoured route for algal biomass conversion. Significant use is already being made of gas from land-fill in the United States, and there are a number of programmes which claim that up to 10 percent of the gas needs in the United States in the next century can be supplied from algae farms or seaweed cultivation.

Methane is also a primary component of wood gas and can be synthesised from it in larger quantities by some pyrolytic processes. Gasification is a favoured route for dry biomass like wood and other materials with a high content of lignin.

The idea of the 'perpetual methane economy' as put forward by Klass (1974)

OVERALL PROSPECTS AND LIMITATIONS

involves the eventual substitution of exhausted natural gas (of which there is already a shortage in the United States) by synthetic natural gas (SNG) provided from renewable carbon sources. It is relatively easy, making asuumptions about annual yields of plant water, and efficiencies for conversion to gas to estimate the land areas and matter quantities that would be required to supply the SNG needs of any country. Though, on reflection, the figures are fearsome in terms of the dislocations and investment that would be involved, for a number of countries it could be a feasible operation provided the overall economics were favourable and, in particular, in common with most solutions, the question of fertiliser requirements were solved.

Alternatively, or additionally, the oceans offer a similar opportunity. For ocean kelp farming schemes as described in chapter 9, it is calculated by Wilcox (1980) that some 40 000 ha (that is an area 40 km long × 10 km wide) in low latitudes could generate up to 620 million m^3 of methane per year. This would be equivalent in energy terms to about half a million tonnes of oil.

A2 METHANOL

It is liquid fuel however, about which there is the greatest concern regarding world supplies. Liquid, transportable, petroleum substitutes are of great interest and gas production is part of the route by which these hydrocarbon fuels, of which methyl alcohol or methanol is the simplest, are synthesised. In particular wood gas or synthesis gas produced by high temperature processes is the first stage in methanol processing and this solution is of interest to countries with very large timber reserves like Canada.

The use of methanol as a large-scale liquid fuel substitution, either as it is or as an additive or stretcher in gasoline, can be taken more seriously because it can also be synthesised from natural gas and coal as well as wood, so unlike other biomass fuel alternatives it is not so constrained by the available quantities of plant matter and the land areas required to produce them. Indeed large-scale synthesis of liquid fuel from coal may well be the necessary step before transition to a renewable biomass feedstock strategy can be developed as the fossil stocks begin to deplete.

Extensive studies in Canada have indicated that there the feedstock resources are adequate to produce the quantities of methanol sufficient to replace a significant proportion of petroleum use in the country, and that large-scale supplies could be reasonably phased into the energy-use scene by the mid-1990s, (Intergroup, 1976). Rather than competing for forestry resources, as conceived, the strategy would enhance the forest industry by improved management practices of winning and profitable utilisation of wastes. Few countries have the amount of forest resources of Canada, and for this reason methanol contributions from biomass can only be a partial solution in the world as a whole. Further it may be considered to be a solution primarily for temperate zones for in sub-tropical and tropical conditions the alternative production of ethanol from plantation crops may have greater attractions. In the case of Brazil both these possibilities are being exploited.

A3 ETHANOL–GASOHOL

Ethanol is ethyl alcohol, produced mainly by the fermentation of sugar and starch-rich feedstocks, but also from predominantly cellulosic material if this can be hydrolysed by acid treatment. However, rather than utilising forest and crop waste, the best prospect for quantity production of ethanol is currently considered to be from field crops like grain, sugar cane and starch-rich roots.

Its use in internal combustion engines either as complete fuel or a gasoline additive has a long history and has been important locally, but until recently it has never competed seriously, or been allowed to compete, with gasoline. Now we see in Brazil a major effort to convert the country's transport to using gasohol, a mixture of ethanol and petrol, and already a substitution of up to 20 percent of current petrol consumption appears to be a feasible short-term target. Gasohol is also being used increasingly in the United States with substantial government encouragement and support and it has been postulated that it could be reasonably used for 30 percent of auto fuel requirements in the mid-1980s. It is argued here, that an ethanol production programme would help solve the dual problems of liquid fuel shortage and Mid-west grain surpluses.

In Brazil large scale production from sugar cane for fuel use started after the 1973 oil supply crises, where previously, just prior to this, the production of ethanol as a chemical feedstock from sugar could not compete with its synthesis from oil. Because of the sharp turn-round in the economic situation and the country's dependence on imported oil – about 80 percent – the National Alcohol Programme was established in 1975 with the supplies of sugar cane for this separated from the normal sugar industry supplies. Other lines of development are underway using manioc (cassava) roots and sorghum which with sugar cane are very well suited to large-scale high-yield production in the tropical environment of Brazil, which also has abundant land and rainfall.

A number of other countries are taking an interest in 'power alcohol' and Zimbabwe is in process of switching entirely to gasohol, importing only regular grade petrol and increasing its octane rating with 14 percent ethanol plus one percent benzol to make the alcohol unpalatable for illicit drinking (denaturation). The ethanol is produced from the countries own sugar cane resources. Already (in 1980) power alcohol projects like this have made molasses scarce and expensive in the world market.

Ethanol, like methanol, has the advantage that internal combustion engines can be easily converted to use it. It is also clean burning and non-polluting. Its disadvantage for transport is that like most substitution fuels it has a lower energy density than gasoline and hence a larger volume must be carried by the vehicle for the same mileage. If freely available it is also open to abuse and small-scale unofficial exploitation of fermentation technology has been rightly hampered by customs and excise policy. However, the technology is highly flexible and viable both for large-scale and home production and as such is a very promising line of development even though there are doubts about the overall economics of production and the fact that large-scale schemes compete with food production. In Brazil for instance it is high government taxation on petrol that makes gasohol competitive at present.

A4 THE HYDROGEN ECONOMY

The idea of this is that liquid hydrogen replaces petroleum and other liquid and gaseous fuels, with water being the universal and virtually unlimited source of the gas. The basic proposition is that the hydrogen is generated electrolytically, but the idea is discussed here because of the alternative (or additional) possibility of biophotolysis (see chapter 9). Effective biophotolysis systems could have an economic and energy budget advantage over electrolytic splitting of water which is energy-intensive. Like most conventional fuel substitutes hydrogen suffers from the disadvantage that larger quantities are needed for the same energy output. It will also require pressurisation and refrigeration if it is to be transported conveniently. Despite this the idea has been treated seriously enough to lead to the design and construction of experimental hydrogen powered vehicles, and for design studies to be carried out for aircraft. Liquid hydrogen is already a major fuel for rocket propulsion, and because of its clean burning and the lack of polluting effects from its use, is likely to become the main fuel as we enter the age of the space shuttles.

The prospect for biophotolysis is exhibited by both bacteria and algae and has also been demonstrated using not only whole-cell plants but chlorophyll and plant enzyme extracts. Maximum efficiency is not likely to exceed 13 percent (Hall *et al.*, 1980) and on this fact it is postulated that if a continuous system with 10 percent efficiency could be established, it would require only 0.1 percent of the Earth's surface to supply the total world energy needs in the form of hydrogen. Similarly, Mitsui and Kamazawa (1977) claim that a hydrogen producing algal culture over an area of 215 km^2 would supply current US hydrogen needs (see chapter 9). However, the current state of the art is that though several biophytolysis systems appear possible, there has been no demonstration yet of a stable long-term working process. Despite its problems, a high demand for hydrogen fuel is certain, even if it does not become a universal fuel. Whether solar-powered biophotolysis generation can supply a significant proportion of this demand in competition with electrolysis possibly using cheap electricity from nuclear sources is one of the more open questions in the whole biomass field.

A5 SOLUTIONS IN PERSPECTIVE

Solar energy, by whatever route it is obtained, is not free. As pointed out by Slesser and Lewis (1979) and others, it cannot be unless the land used for a production scheme is free and the capital employed in it has no cost. Land used for energy plantations must therefore give a greater return than it would if it were to be used for other purposes. Little land anywhere, except perhaps some very rough terrain, desert and arctic waste has negligible value, and certainly no land that could grow a crop of any sort. Sea areas or salt water lagoons and lakes may appear promising in this respect. However, the more difficult the site the increase in the engineering costs more than offsets the gain from low site values. This bears on the second point that finance has to be found for the infrastructure of a biomass fuel producing scheme and as well as paying back the capital and the interest on the capital, the scheme must earn enough to cover depreciation and replacement. Obviously fuel from biomass schemes must be economic in overall terms, but in matters of energy balance the situation is not necessarily so straightforward.

Biomass fuels are only one of a number of alternative energy possibilities, and in assessing their viability much is made of the energy accounting approach. For instance, the production from plant material of ethyl alcohol for say beverage and chemical uses is generally claimed to absorb more energy than is provided in the alcohol output. Though large-scale fuel alcohol plants aim to achieve a positive energy gain, this tends to be small. However, a too rigid insistence on energy gain in the process ignores the quality of the energy, or rather the nature of the fuel produced. For instance a liquid fuel produced by a process that is a net consumer of energy is perfectly acceptable if the process energy is provided, say, from crop waste, as liquid fuels are premium fuels which because of their energy density, flexibility and transportability are worth so much more than equivalent quantities in energy terms of the original biomass.

One way of comparing the viability of different fuel-energy approaches is to compare the cost of producing electricity by the various processes. For example ERDA quoted the following comparative costs in 1976 (Hammond and Metz, 1977):

Method	Cost factor in relative to conventional electricity generation
Wind	1–2
Biomass	2–4
Ocean thermal generation	4–5
Direct solar energy capture (heliostat systems)	5–10
Photo-voltaic systems	20–40 (but expected to fall rapidly)

This approach probably undervalues energy from biomass as its most effective

OVERALL PROSPECTS AND LIMITATIONS

contributions are probably in the supply of liquid transportable fuels for petroleum substitutes and gas for local consumption, in both cases purposes for which electricity produced and supplied by conventional means may not be appropriate. Biomass is therefore, best considered in terms of its fuel product rather than by its gross, or even net energy values. Even if, in some decades, nuclear fusion power is able to produce unlimited electricity at low cost, it may well make the generation of fuels from biomass more economic rather than out-compete them.

On the other hand, any 'alternative' fuels are considered to be more and more economic to produce and use as the cost of conventional fuels, (particularly oil), and energy overall rises. The matter, however, is not as straightforward as this as the 'basic' fuel costs will add considerably to the capital costs of the biomass schemes. For this reason any additional value arising in connection with the schemes — like waste treatment — will give a valuable boost to the value of any proposition.

In the final analysis any significant contributions to the world energy supply from biomass face two fundamental problems:

(i) In the case of plantation type schemes, they compete with food production and are limited by land availability and by available nutrient supplies.

(ii) In the case of waste-utilising schemes they are even more absolutely limited by the quantities of waste actually available.

B National programmes

Even though much of the biomass-energy technology has a long history, the OPEC petrol price increase of 1973 was undoubtedly the boost for the huge increase in the subject, and also for the beginnings of the formulation of most bio-energy development plans made against the realisation of the limits of supply of the main fossil fuels. Brazil clearly leads the way in its national programme of practical realisation of fuel substitution but many if, not most, countries are carrying out policies of developing bio-fuels at different scales or have at least established co-ordinated research and development programmes.

B1 BRAZIL

Here, as we have seen, a practical programme of large-scale liquid fuel production is well underway. The initial concentration is on liquid fuel substitution by ethanol based on established fermentation technology using cane and manioc (cassava) with interest developing in other starchy crops like soya beans. However, an additional alternative strategy using wood and dry crop waste to generate methanol by pyrolytic processes is also being heavily invested in. The 'Proalcool' programme of substituting gasoline by a 20 percent of gasoline/alcohol 'gasohol' mixture shows every sign of meeting its short-term targets and ethanol production is planned to rise from 3.8 million m^3 (B/l) in 1979 to 10.7 million m^3 in 1985 from about 250 new distillery complexes. This will account for 20 percent of liquid fuel requirements in 1985 and will require 2.6 million ha of agricultural land to provide the necessary energy crops. This is about 20 percent of the present agricultural land and will represent a total programme investment of the order of $5 billion. To adapt to the changing fuel policy, some 900 000 new vehicles designed to run on proalcool or pure ethanol are expected to be in operation in 1985.

The main justification for this programme is the cost saving on imported oil. Though Brazil is energy-rich in hydro-electricity and coal and even has some petroleum of its own, with a huge land area to be developed the demand for flexible transport based on liquid fuels is the key factor for the future.

B2 THE UNITED STATES

Of all the programmes of research and development, that in the United States is the most extensive and wide ranging and it is difficult to identify any particular avenue of investigation that is not being explored by a substantial effort. The United States is also the home of many of the more revolutionary concepts and propositions. Federal research budgets at the beginning of 1980s are of the order of $60 million and showing every sign of increasing.

At the practical level a number of technologies are being implemented, though there is no single large-scale strategy in prospect equivalent in importance to the Brazilian proalcool programme. However, the largest concentration of efforts is on the use of forest biomass, particularly to meet the energy needs of the forestry industry but also extending, in forest states, to the generation of electricity. Largely this is by extension and improvement of the already long established practices of combustion and to some extent this can be seen as a 'natural' tendency instigated by the increasing cost of other fuels coming to out-weigh the disadvantages of wood and wood waste use.

The use of municipal solid waste is also a major area of effort with the technologies overlapping those of wood waste use. Here, the additional, associated, benefits of waste recovery and recycling, disposal and environmental protection are important, even the main considerations.

In such a varied field of prospects there are naturally numerous lobbies for particular solutions and gasohol, which was originated in the 1920s and 1930s based on the use of grain surplus and was killed off then by the oil companies, is again attracting interest and action. Moves towards large-scale programmes are underway with federal encouragement for investment in new distillation plant. Ethanol, however, is a valuable chemical feedstock and if, as seems likely, it can now be produced more cheaply from starch crops rather than petroleum feedstocks, much may be used in place of oil to synthesise other chemicals rather than as fuel. The proposed production in 1981 of 500 million gallons (1.89 Mm^3) is a six-fold increase over the 1979 production, and there is a target for 1985 of

OVERALL PROSPECTS AND LIMITATIONS

2 billion gallons (7.56 Mm^3). The raw materials for this will be maize surpluses, with a possible move towards specifically grown sweet sorghum, both crops at present being largely produced for animal feed. These quantities of alcohol however, represent a much smaller fraction of the national liquid fuel demand than they do for the Brazilian national programme and because of much higher fuel demand in the United States the ceiling of land available to grow the biomass feedstock is much more limiting particularly as virgin land which can be converted to fuel crops is much more limited.

B3 CANADA

Because of its huge forestry resources Canada has possibly the largest stock of biomass per capita of any country in the world and is developing a concerted national programme to exploit it with a target of increasing its proportion of renewable energy (including solar, hydro etc.), from 3.5 percent to 10 percent by the year 2000 (Overend, 1978).

Overwhelmingly the biomass is mostly wood and as 7.5 percent of the national energy consumption is in the forest industries, this is obviously the prime target area for conversion to wood-fuel use, both by direct combustion and the generation of wood gas and methanol. Agricultural residues are considered nowhere near as attractive a prospect. The Energy from Forest Programme (ENFOR) has as its immediate objective the establishment of the basis to permit replacement of non-renewable fuel up to 8 percent. Investment in this in 1979 was of the order of $30 million.

B4 EUROPE

Most European countries have significant biomass programmes of their own, like the French 'Alter Programme' which aims to supply 10 percent of energy requirements from alternative sources including biomass by the year 2000, but in addition to these the Commission of the European Communities are also carrying out a co-ordinated programme of biomass energy development as part of the larger solar energy programme of the Directorate-General for Research, Science and Education. However, early expenditure on biomass research was only 10 percent of the total solar energy spending, but this has risen to 16 percent in the second four-year programme from 1980 to 1984 — a total of $10 million or so with an additional $3.8 million to be spent on research into photosynthesis. The early programme was very comprehensive and varied lines of research are being followed in the second programme. In this, however, a primary objective is the development of practical methods of gasification and methanol production from wood and dry crop wastes. This is seen in the context of a series of phased development priorities of:

1. 1975–1985: Utilisation of residues to produce energy for local needs.
2. 1980–1995: Development of energy crops on available land without marked modifications to the current agricultural situation.
3. 1985–2000: Development of energy plantations.

Phases 2 and 3 are intended to produce increasing quantities of liquid fuel. Following a path of major reorganisations of under-developed land it has been estimated that up to 20 percent of the total fuel requirements required by the Community in 1985 could be supplied from biomass. This however, possibly represents the practical limit achievable from terrestrial sources.

One practical outcome of the programme is that a large-scale short rotation forestry development scheme is to be set up in Ireland as a demonstration programme funded by another part of the European Commission.

B5 DEVELOPING COUNTRIES

Of these, the most impressive achievements are those of China with its domestic and farm scale biogas programme which became really effective in 1975 with a breakthrough in digester design. This approach is ideally suited to the intensive agricultural situation and the 7 million or more digesters in operation are considered to be a major factor in the improvement of the rural standard of living (see Van Buren and Pyle, 1979).

With a similar situation in India biogas is also seen as the most important line of development. Despite some religious objections about the association of human and animal waste with the use of gas for cooking a programme to produce 'Gobar' gas units is under way with about 40 000 units in operation in 1980. (Gary et al., 1980). With a similar climate, the new nation of Zimbabwe is following the same path as that of Brazil. In 1980 the first 'gasohol' began to come into use, composed of 14 percent ethanol and one percent benzol* added to 'regular' grade fuel. Capacity to produce enough ethanol for a complete 14 percent substitution is being installed and for an expansion to a 20 percent substitution in petrol and a 15 percent substitution in diesel fuel. The biomass feedstocks source is initially sugar cane with the ethanol plant integrated with a sugar mill complex. However, sorghum and cassava are being investigated as sources. It has been speculated that Zimbabwe has enough land and a low enough population and related energy demand to become self-sufficient in liquid fuels through ethanol.

Biomass fuel programmes are increasingly appearing as elements in international aid programmes. Though there is resistance to these in some quarters on the grounds that they are 'second class' technologies the potential for energy and cost saving in individual development schemes is likely to ensure increasing adoption of the methods as components of package solutions, whereas distinct bio-energy strategy programmes are somewhat less likely.

B5 INSTITUTIONAL AND POLITICAL CONSTRAINTS

The objection in some quarters in developing countries to alternative technology represents one form of institutional constraint to biomass energy development, additional to the basic opportunity, technological and economic constraints. The

---

*Benzol improves the running of the engine as well as making the ethanol undrinkable

OVERALL PROSPECTS AND LIMITATIONS

killing-off of the original gasohol programmes in the United States by the oil companies is an example of another, though it had a commercial motive, whereas the objection from the developing countries has a political motive. Though fossil fuel short-falls and government pressures now reduce the dangers of commercial pressures (as opposed to economic facts) hindering biomass programmes, political attitudes in some developing countries could prove to be major obstacles which will only be removed where the advantages of using biomass on a large-scale (though not necessarily in large-scale units) have been demonstrated.

In a specific case, the constraints imposed by the conservatism of the existing establishments on the use of municipal and solid waste are given in a report by the United States Department of Energy (Gordian Associates, 1977). Here, the principal objection to the adaption of the methods is seen to be a fundamental aversion to risk. That this is relatable to the perceived state of the art of the technologies is illustrated by the actual degree of acceptance of the individual possibilities, with electricity generation from combustion being most applied, some use of steam generation, but little acceptance of solid fuel forming (RDSF), and use and operational pyrolysis being negligible. This equates exactly with the degree of technical development for the various options. Not surprisingly the key factor in the introduction of the methods is seen to be in financial encouragement by state or federal authorities, not necessarily by direct subsidy but by appropriate legislation and forms of tax relief, for instance the relaxation of petrol tax on gasohol in the United States in 1978. Clearly, political attitudes can be influenced by the demonstrated feasibility of the technologies and these themselves can be affected by broader political issues. For instance the overriding factor of market acceptance will be governed as much by the relative cost of biomass relative to fossil fuels, which though depending ultimately on world supplies, can be considerably influenced by national resources and economic policies. With the world energy market certainly liable to somewhat unpredictable dislocations in the next few decades at least, the exploration of alternative biomass options is an important political objective.

The ways in which governments can assist biomass development are; the normal process of funding basic science and high risk development research; the provision of flexible legislation to take account of detailed requirements of particular solutions; certification of the systems developed and encouragement of programmes for their adoption; and making available land, water and other necessary resources. All these are against the background of perceived national good which must also take account of other factors like competition with food production.

Political factors may, however, distract programmes for the benefit of sectional interest, and a willingness to believe in magic solutions by some can further complicate the situation. No mention has been made so far of the alternative 'Biomass Society' as a complete philosophy. Anybody who has a garden is aware of the problems of labour-intensive agriculture and a world predominantly composed of self-sufficient communities is only likely to come about as the result of catastrophy. The setting up of energy-self-sufficient facilities is, however, a valuable research approach even though living in them as a serious lifestyle must be thought of as on a par with nudist colonies.

In the light of the more realistic projections for biomass use, which do not exceed 10 percent of total energy requirement for any country by the year 2000, the globally attractive features of biomass energy, *viz.* (*i*) renewable sources, (*ii*) pollution from low sulphur fuels, including reduction of the 'acid rain' problem, (*iii*) reduced contribution to the atmospheric carbon dioxide build-up and (*iv*) nutrient recycling, become relatively minor considerations. However, 20 percent of a nation's liquid fuel requirements is something to be reckoned with and there is plenty of evidence that biomass energy will be a significant force in the future energy mix. The moves in this direction, therefore, must not be seen as one method of saving the world, but rather part of a general progress towards greater order and efficiency in the way that resources are utilised.

The state of the art varies considerably with the many different prospects but the fact that at one end of the scale the techniques are proven and of longstanding indicates that the pattern of development will be a steady climb up the ladder of possibilities. The basic constraint of the low flux density of received solar radiation cannot be avoided and breakthoughs like nuclear fusion cannot be expected. However, there is scope for individual refinements in numerous areas, giving many fruitful fields of research. Perhaps the most significant advances will occur in the areas of plant productivity — which is what biomass energy is basically about — with self-fertilisation by implanted nitrogen-fixing characteristics, breeding of plant types with high energy densities and high growth rates, and the development of mass reproduction and species characteristic control methods being some of the most important. On the systems side, the integration of biomass energy production with waste disposal and environmental improvement, energy and materials recycling, and chemicals and food production are not only valuable goals in themselves, but also probably the paths by which many of the prospects will be developed.

At the time of writing, although it is still recognised as the main problem for the future (and present), the anxious views and reactions of the early 1970s to the energy crisis are less in evidence. The idea, based on simple extrapolations that all petroleum will be used up by the year 2020, is no longer current. This is due to a number of factors including; a realisation that readily recoverable reserves are probably larger than thought; that higher recovery rates from existing fields and exploitation of more difficult sources very much increase the potential reserves and that the forcing up of prices by demand and scarcity will make the exploitation of these viable. On the other hand even nuclear fusion, if it is ever achieved, is unlikely to be a magic solution solving all problems.

OVERALL PROSPECTS AND LIMITATIONS

The passing of the age of cheap energy will also increase the feasibility of various alternative energy possibilities. We have seen that energy from biomass can, like most other alternative energy sources, like wind and wave energy, only contribute a fraction of the energy requirements of advanced industrial nations. Individually the opportunities are closely tied to the sources of their raw materials. Most importantly recovering energy from biomass will serve to take some of the pressure off primary fossil fuel requirements and increase options, which in a complex society is always a good thing.

Units and abbreviations

bar	10^5 newtons per m^2 or approximately 1 atmosphere
Btu	British thermal unit (1.055 KJ)
Btu/scf	British thermal unit per standard cubic foot (0.37 KJ/m^3)
cal	calorie (4.184 J)
dt	Dry tonne
EEC	European Economic Community
EJ	exajoule (joules × 10^{18}, approximately equivalent to 1 quad)
GJ	gigajoule (joules × 10^9)
g	gram
g/kg	grams per kilogram
GPP	Gross Primary Product
ha	hectare (10 000 m^2)
J	joule (1 watt/second)
kg	kilogram (2.2 lb)
kcal	kilocalorie (calories × 10^3)
lb	pound (0.45 kg)
ly	Langley (1 cal/cm^2 or 4.184 J/cm^2)
LNG	Liquified Natural Gas
M	million
MJ	megajoule (joules × 10^6)
MSW	Municipal Solid Waste
Mt	megatonne (tonne × 10^6)
Mtoe	Million tonne oil equivalent (44 × 10^6 GJ)
NPP	Net Primary Product
N/t	nitrogen content per tonne
odt	oven-dry tonne
pa	per annum
PGA	Phosphoglyceric acid
psig	pounds per square inch (70 g/cm^2) gauge. The excess of pressure over 1 atmosphere
QAM	Quarter Acre Module
Quad	10^{15} Btu or approximately 10^{18} J
RDSF	Refuse-Derived Solid Fuel
SNG	Synthetic Natural Gas
TJ	terajoule (joules × 10^{12})
t	tonne (1000 kg)
W	watt (1 joule/second)
Wh	watt/hour (3.6 KJ)
yr	year

References

NOTE

Some of the recent research work that has been drawn upon for this book has either not yet been published at the time of writing, or only in relatively inaccessible documents such as specialist symposia proceedings and limited issues by research agencies.

As far as possible we have avoided references that are too obscure, if necessary quoting more general journal articles which represent synopses of the work by the authors involved. In some cases there has been no option. However, although many of the references may not be available in university libraries they can be obtained through central scientific libraries and in the case of some government publications through direct application to the departments or bodies concerned.

Chapter 1

Boardman, N. K. and Larkum, A. W. D. (1975). Biological conversion of solar energy. *In* "Solar Energy" (Messel, H. and Butler, S. T., eds). Pergamon, London

Frissel, M. J., Van Goar, C. P., de Hoop, D. W. and Olie, J. (1978). "Feasibility of Short Rotation Forestry for Energy Purposes". Project E Study Report. E.E.C., Brussels. (Quoted in Palz and Chartier, 1980)

Hogg, W. H. (1971). Regional and social environments. *In* "Potential Crop Production" (Wareing, P. F. and Cooper, J. P., eds). Heinemann, London

Long, G. (1977). "Solar Energy: Its Potential Contribution within the U.K.". Energy Paper No. 16. HMSO, London

Loomis, R. G. and Gerakis, P. A. (1975). Productivity of agricultural ecosystems. *In* "Photosynthesis of Plants as a Factor of Primary Production". Academic Press, London and New York

Merril, R. (1974). "Energy Primer". Portola Institute, California

Newbold, P. J. (1971). Comparative production of ecosystems. *In* "Potential Crop Production" (Wareing, P. F. and Cooper, J. P., eds). Heinemann, London

North, W. J. (1977). The ocean setting as a site for biomass production. *In* "Fuel from Biomass". University of Illinois, Urbana

Palz, W. and Chartier, P. (eds). (1980). "Energy from Biomass in Europe". Applied Science Publishers, London

Pearsall, W. H. and Gorham, E. (1956). Production ecology, I. Standing crops of natural vegetation. *Oikos* 7(11), 193–201

Penman, H. L. (1971). Water as a factor in productivity. *In* "Potential Crop Production" (Wareing, P. F. and Cooper, J. P., eds). Heinemann, London

Round, F. E. (1966). "The Biology of the Algae". Edward Arnold, London

Ryther, J. H., Lapointe, B. E., Stenberg, R. W. and Williams, L. D. (1977). Cultivation of seaweeds as a biomass source for energy. *In* "Fuels from Biomass". University of Illinois, Urbana

Shelef, G., Moraine, R., Meydan, A. and Sandbank, E. (1976). Combined algal production–waste treatment and reclamation systems. *In* "Microbial Energy Conversion". Erich Goltze, Gottingen

Schneider, T. R. (1973). The efficiency of photosynthesis as a solar energy converter. "Energy Conservation", Vol. 13, Pt. 3, p. 77. Pergamon, Oxford

Soeder, C. J. (1976). Primary production of biomass in freshwater with respect to microbial energy conversion. *In* "Microbial Energy Conversion". Erich Goltze, Gottingen

Thom, G. N. (1971). Physiological factors limiting the yield of arable crops. *In* "Potential Crop Production" (Wareing P. F. and Cooper, J. P., eds). Heinemann, London

Westlake, D. F. (1963). Comparison of plant productivity. *Biol. Review* (Cambridge Phil. Soc.), No. 38, 385

Chapter 2

Alich, J. A. and Inman, R. E. (1976). "An Evaluation of the Use of Agricultural Residues as an Energy Feedstock", Vol. 1. Stanford Research Institute, Project 3520. Menlo Park, California, USA

Baldwin, E. (1952). "Dynamic Aspects of Biochemistry". Cambridge University Press, Cambridge

Bliss, C. and Blake, D. O. (1977). "Silvicultural Biomass Farms, Conversion Processes and Costs", Vol. 5. The Mitre Corporation, Metrek Division, USA

Brooks, B. T. (1942). Petroleum research and wars. *Ind. Eng. Chem.* 34, 798

Converse, A. D., Grethlein, H. E., Karandiker, S. and Kuhrtz, S. (1971). "A Laboratory Study and Economic Analysis for the Acid Hydrolysis of Cellulose in Refuse to Sugar and its Fermentation to Alcohol". Final Report to the PHS (US). Grant No. US-00597-02, Thayer School of Engineering, Dartmouth College, Hanover, USA

Crane, T. M., Ruger, H. N. and Bridges, D. W. (Barber Colman Co.) (1975). "Production of Gaseous Fuel by Pyrolysis of Municipal Solid Waste". Pilot plant tests on simulated MSW 1400°F molten lead in moving hearth. N75–24105/9 ST

Douglas, E., Webb, M. and Dabourn, G. R. (1974). The pyrolysis of waste and product assessment. Paper given to the Symposium on Treatment and Recycling of Solid Wastes, Manchester. Institute of Solid Wastes Management, Warren Spring Laboratory, Herts., UK

Foo, E. L. and Heden, C. G. (1977). Is biocatalytic production of methanol a practical proposition? *In* "Microbial Energy Conversion" (Schlegel, H. G. and Barnea, J., eds), p. 267. Erich Goltze, Gottingen

Francis, W. (1961). "Coal". 2nd Edn. Edward Arnold, London

Fry, L. J. (1975). "Practical Building of Methane Power Plants for Rural Energy Development". D. A. Unox, Andover, Hants.

Hansford, R. J. (1974). Other possibilities for processing straw. *In* "Report on Straw Utilization Conference" (Staniforth, A. R., ed.), p. 26. Ministry of Agriculture, Fisheries and Food

Hayes, T. D., Jewell, W. J., Dell'Orto, S., Fanfoni, K. J., Leuschner, A. P. and Sherman, D. F. (1979). "Anaerobic Digestion of Cattle Manure" (Stafford, D. A. and Wheatley, B. E., eds). A. D. Scientific Press, UK

Isman, B. – quoted by Wheatley, B. (1979). "Anaerobic Digestion on Farms – The European Scene". Contribution to ADAS Conference on Anaerobic Digestion, Cardington, Beds.

REFERENCES

Jackson, E. A. (1976). Brazil's national alcohol programme. *Process Biochem.* 11, 29–30

Knight, J. A. (1976). "Pyrolysis of Fine Sawdust". 172nd American Chemical Society National Meeting. San Fancisco, USA

Lewis, F. M. and Ablow, C. M. (1976). Pyrolysis of biomass. *In* "Capturing the Sun through Bioconversion", p. 34. Washington Center for Metropolitan Studies, Washington D.C.

Lucas, J. (1979). Energy from biomass by gasification of straw. Proceedings of the Third Co-ordination Meeting of Contractors, "Energy from Biomass" (Project E), p. 5.2.1. E.E.C. Solar Energy R and D Programme, Toarmina, Italy

McCann, D. J. and Saddler, H. D. W. (1976). Utilization of cereal straw; a scenario evaluation. *J. Aust. Inst. Agric. Sci.* March, 41–47

McCarty, P. L., Ross, E. and McKinney, T. (1961). Salt toxicity in anaerobic digestion. *J. Water Pollution Control Fed.* 33, 399

Potter, O. E. and Keogh, A. J. (1979). Cheaper power from high moisture content brown coals, *Part I*; System description and preliminary costing. *J. Inst. of Energy* September; 143: *Part II*; Feasibility and development, as above, 145

Ratledge, C. (1976). Microbial production of oils and fats. *In* "Food from Waste" (Birch, G. G., Parker, K. J. and Worgan, J. T., eds), p. 98. Applied Science Publishers, London

Rijkens, B. A. (1979). Methane and compost from straw. Proceedings of the Third Co-ordination Meeting of Contractors, "Energy from Biomass" (Project E), p. 2.3.1. E.E.C. Solar Energy R and D Programme, Toarmina, Italy

San Pietro, A. (1977). Hydrogen formation from water by photosynthesis and artificial systems. *In* "Microbial Energy Conversion" (Schlegel, H. G. and Barnea, J., eds), p. 217. Erich Goltze, Gottingen

Smith, I. E. and Hansford, R. J. (1976). Densification – a pre-requisite to the industrial use of straw. *In* "Report on Straw Utilization Conference" (Staniforth, A. R., ed.), p. 26. UK Ministry of Agriculture, Fisheries and Food

Smithson, G. R. (1977). Utilization of energy from organic wastes through fluidized bed combustion. *In* "Fuels from Waste" (Anderson, L. L. and Tillman, D. A., eds), p. 195. Academic Press, London and New York

Soeth, S. (1973). Low molecular weight products from P/E wastes liquid stirred reactor, molten salt as external heat transfer medium. *Chem. Ind.* 12, 557–559

Strehler, A. (1979). Proceedings of the Third Co-ordination Meeting of Contractors, "Energy from Biomass" (Project E), p. 1.2.1. E.E.C. Solar Energy R and D Programme, Toarmina, Italy

Thauer, R. K. (1977). Limitation of microbial hydrogen formation via fermentation. *In* "Microbial Energy Conversion" (Schlegel, H. G. and Barnea, J., eds), p. 201. Erich Goltze, Gottingen

Tornabene, T. G. (1977). Microbial formation of hydrocarbons. *In* "Microbial Energy Conversion" (Schlegel, H. G. and Barnea, J., eds), p. 281, Erich Goltze, Gottingen

Varel, K. H., Isaacson, H. R. and Bryant, M. P. (1977). *Appl. & Envir. Microbiol.* 33(2), 298

Chapter 3

Ader, G. and Buck, F. R. (1979). "Organic Wastes as an Energy Source". Study for Energy Technology Support Unit, Harwell, UK

Arnason, J. and Magne Mo. (1977). Ammonia treatment of straw. *In* "Report on Straw Utilization Conference" (Staniforth, A. E., ed), pp. 25–31. UK Ministry of Agriculture, Fisheries and Food

Bowerman, P. (1977). Feeding trials with NaOH treated straw at ADAS experimental husbandry farms. *In* "Report on Straw Utilization Conference" (Staniforth, A. R., ed.), pp. 36–38. UK Ministry of Agriculture Fisheries and Food

Chartier, P. *et al.* (1978). Production of cereal straw as a source of fuel and related products. Proceedings of the Second Co-ordination Meeting of Contractors, "Energy from Biomass" (Project E). E.E.C. Solar Energy R and D Programme, Brussels

Converse, A. D., Grethlein, H. E., Karandiker, S. and Kuhrtz, S. (1971). "A Laboratory Study and Economic Analysis for the Acid Hydrolysis of Cellulose in Refuse to Sugar and its Fermentation to Alcohol". Final Report to the PHS (US). Grant No. US-00597-02. Thayer School of Engineering, Dartmouth College, Hanover, USA

Coxworth, E. (1978). Some aspects of straw utilization in Western Canada. *In* "Report on Straw Utilization Conference" (Staniforth, A. R., ed.), pp. 126–128. UK Ministry of Agriculture, Fisheries and Food

Francis, G. H. (1976). Use of vegetable and arable by-products. *In* "By-Products and Wastes in Animal Feeding" (Orskov, E. R., ed.), Occasional Publication No. 3. British Society of Animal Production

Greenhalgh, J. F. D. and Pirie, R. (1977). Alkali treatment of straw: a summary of experiments made at the Rowett Research Institute. *In* "Report on Straw Utilization Conference" (Staniforth, A. R., ed.), pp. 23–24. UK Ministry of Agriculture, Fisheries and Food

Hansford, R. J. (1974). Other possibilities for processing straw. *In* "Report on Straw Utilization Conference" (Staniforth, A. R., ed.), p. 26. UK Ministry of Agriculture, Fisheries and Food

Hughes, R. (1977). Fate of cereal straw – 1976. *In* "Report on Straw Utilization Conference" (Staniforth, A. R., ed.), pp. 2–5. UK Ministry of Agriculture, Fisheries and Food

Johnston, A. E. (1978). Some beneficial effects on crop yields from incorporating straw into the plough layer. *In* "Report on Straw Utilization Conference" (Staniforth, A. R., ed.). UK Ministry of Agriculture, Fisheries and Food

Loll, U. (1977). Engineering, operation and economics of bio-digesters. *In* "Microbial Energy Conversion" (Schlegel, H. G. and Barnea, J., eds), p. 361 Erich Goltze, Gottingen

Lucas, N. G. (1978). Whole crop harvesting: 1, 2 and 3. *Power Farming* August, September, October

McCann, D. J. and Saddler, H. D. W. (1976). Utilization of cereal straw; a scenario evaluation. *J. Aust. Inst. Agric. Sci.* March, 41–47

Meriaux, S., Begon, J. C., Gachon, L., Hutter, W., Juste, C, and Cochin, B. (1977). "Production de Céréales et de Pailles de Maïs a des Fins Énergétiques". Contract D53 76 ESF, I.N.R.A., Paris

Monteith, J. L. (1977). Climate and the efficiency of crop production in Britain.

REFERENCES

In "Agricultural Efficiency", p. 203. The Royal Society, London. Also *Phil. Trans. R. Soc. Lond.* B 281, 277–294

Morris, J., Smith, D. L. O. and Radley, R. W. (1977). "A Cost Analysis of Alternative Systems of Straw Procurement". Occasional Paper No. 6. National College of Agricultural Engineering, Beds.

Moteurs Duvant (1979). Gas generation of low BTU fuel gas from agricultural residues. Proceedings of the Third Co-ordination Meeting of Contractors, "Energy from Biomass" (Project E), p. 5.4.1. E.E.C. Solar Energy R and D Programme, Toarmina, Italy

Nuttall, M. (1978). Making and feeding sugar beet top silage to beef cattle. *In* "Animal Feeds from Sugar Beet Crops". UK Ministry of Agriculture, Fisheries and Food

Nuttall, M. (1979). Beet tops for cattle feeding. *Sugar Beet Review* 47(2), 25

Palz, W. and Chartier, P. (eds). (1980). "Energy from Biomass in Europe". Applied Science Publishers, London

Pedersen, T. T. (1979). Use of straw for heating purposes. Proceedings of the Third Co-ordination Meeting of Contractors, "Energy from Biomass" (Project E), p. 1.3.1. E.E.C. Solar Energy R and D Programme, Toarmina, Italy

Plom, A. (1975). "A Cost/Benefit Analysis of Selected Alternative Straw Disposal Methods". Study report submitted to National College of Agricultural Engineering, Beds.

Porteus, A. (1976). "Towards a Profitable Means of Waste Disposal", ASME Paper No. 67-WA/PID-2. Presented at Winter Annual Meeting and General Systems Exposition, pp. 1–17. American Society of Mechanical Engineers, Pittsburgh, Pa.

Rexen, F. (1976). Straw as an industrial raw material. *In* "Solar Energy in Agriculture", p. 38. International Solar Energy Society, London

Rijkens, B. A. (1977). Some possibilities for multiple uses for staw. *In* "Report on Straw Utilization Conference" (Staniforth, A. R., ed.), pp. 64–68. UK Ministry of Agriculture, Fisheries and Food

Robb, J. and Evans, P. J. (1976). Recycling cereal straw. *In* "Solar Energy in Agriculture", p. 33. International Solar Energy Society, London

Russell, E. W. (1977). The role of organic matter in soil fertility. *In* "Agricultural Efficiency", p. 135. The Royal Society, London. Also *Phil. Trans. R. Soc. Lond.* B 281, 209–219.

Sachetto, J. P. (1977). A project for producing chemicals from straw. *In* "Report on Straw Utilization conference" (Staniforth, A. R., ed.) pp. 71–72. UK Ministry of Agriculture, Fisheries and Food

Spedding, C. R. W. and Walsingham, J. (1978). The potential of photosynthetically-produced organic matter as an energy-feedstock. Proceedings of the Second Co-ordination Meeting of Contractors, "Energy from Biomass" (Project E). E.E.C. Solar R and D Programme, Brussels

Spedding, C. R. W. and Walsingham, J. (1979). The potential of photosynthetically-produced organic matter as an energy feedstock. Proceedings of the Third Co-ordination Meeting of Contractors, "Energy from Biomass" (Project E), p. 6.2.1. E.E.C. Solar R and D Programme, Toarmina, Italy

Stanford Research Institute (1976). "A Preliminary Analysis of the Potential Feasibility of Utilizing Agricultural Residues in the Sutter County, California Area, for Production of Energy". Contract E (04–3) 115, SRI Project 5093

Triolo, L. (1977). Straw pulping for paper in Italy. *In* "Report on Straw Utilization Conference" (Staniforth, A. R., ed.), pp. 54–55. UK Ministry of Agriculture, Fisheries and Food

Truman, A. B. (1974). The utilization of straw for paper-making. *In* "Report on Straw Utilization Conference" (Staniforth, A. R., ed.), pp. 13–17. UK Ministry of Agriculture, Fisheries and Food

University of California (1976). "Technological Assessment of the Utilization of Rice Straw Residue from the California Scaramento Valley for on-Farm Power Generation" (Horsefield, B. and Williams, R. O.). Report No. ERDA/USDA–19464/76/FR–1. Department of Agricultural Engineering, Davis, California

Vahlberg, C. (1978). The Swedish whole crop harvesting project. *In* "Report on Straw Utilization Conference" (Staniforth, A. R., ed.) pp. 11–14. UK Ministry of Agriculture, Fisheries and Food

Widdowson, F. V. (1974). Results from experiments measuring the residues of nitrogen fertilizer given for sugar beet, and of ploughed-in sugar beet tops, on the yield of following barley. *J. Agric. Sci. Camb.* 83, 415–421

Wilkinson, J. M. (1979). Recent developments in chemical treatment of straw for use as feed for livestrock. *In* "Report on Straw Utilization Conference" (Staniforth, A. R., ed.), pp. 17–21. UK Ministry of Agriculture, Fisheries and Food

Wilson, P. N. and Brigstocke, T. (1977). The use of NaOH treated straw (NIS) in rations for dairy cattle – an interim report of further recent work. *In* "Report on Straw Utilization Conference" (Staniforth, A. R., ed.) pp. 39–41. UK Ministry of Agriculture, Fisheries and Food

Wilton, B. (1979). Further development of a straw-fired furnace, to be used initially in conjunction with a crop drying plant. Proceedings of the Third Co-ordination Meeting of Contractors, "Energy from Biomass" (Project E), p. 1.4.1. E.E.C. Solar Energy R and D Programme, Toarmina, Italy

Chapter 4

Ader, G. and Buck, F. R. (1979). "Organic Wastes as an Energy Source". Study undertaken for Energy Technology Support Unit, Harwell by Ader Associates

Forster, C. F. and Jones, J. C. (1976). The bioplex concept. *In* "Food from Waste" (Birch, G. G., Parker, K. J. and Worgan, J. T., eds), p. 278. Applied Science Publishers, London

Green Europe (1978). "Danish Bacon: No End Yet to Expansion Boom". No. 144, p. 23. Agra Europe (London) Ltd

Isman, B. – quoted by Wheatley, B. (1978). "Anaerobic Digestion on Farms – The European Scene". Contribution to ADAS Conference on Anaerobic Digestion, Cardington, Beds.

Loehr, R. C., Prakasam, T. B. S., Skinath, E. G. and Joo, Y. D. (1973). "Development and Demonstration of Nutrient Removal from Animal Wastes". Projects 13040 DPA and 13040 DDG. The Office of Research and Monitoring Environmental Protection Agency, Washington, D.C.

Pain, B. F., Hepherd, R. Q. and Pittman, R. J. (1978). Factors affecting the performances of four slurry separating machines. *J. Agric. Engin. Res.* 23, 231–242

Palz, W. and Chartier, P. (eds). (1980). "Energy from Biomass in Europe". Applied Science Publishers, London

Plaskett, L. G. (1980a). The relationship of biotechnology to biological husbandry. *In* "An Agriculture for the Future". Proceedings of a Symposium of the International Institute of Biological Husbandry, Wye, Kent

Plaskett, L. G. (1980b). "The Generation of Gases from Biomass by Anaerobic

REFERENCES

Digestion and their Practical Utilization as Fuel". Study undertaken for Energy Technology Support Unit, Harwell by Biotechnical Processes Ltd

Robertson, A. M. (1977). "Farm Wastes Handbook". Scottish Farm Buildings Investigation Unit

Taiganides, E. T. and Hazen, T. E. (1966). Properties of farm animal excreta. *Trans Amer. Soc. Agric. Engin.* **9**(3), 374–376

Wilkinson, J. M. (1978). The use of animal excreta as feeds for livestock. *In* "By-Products and Wastes in Animal Feeding". British Society of Animal Production

Yeck, R. G., Smith, L. W. and Culvert, C. C. (1975). Recovery of nutrients from animal wastes — an overview of existing options and potentials for use in feed. *In* "Managing Livestock Wastes". Proceedings of the Third International Symposium on Livestock Wastes, pp. 192–196. American Society of Agricultural Engineers

Chapter 5

Aston, B. C. (1971). Potash in agriculture, III. *N.Z. J. Agric.* **14**, 440–447

Austin, J. E. (1974). "Studies on the Growth and Yields of Annual Forage Crops Grown during the late Summer Period". PhD Thesis in Agriculture. University of Reading

Burton, G. W., Jackson, J. E. and Knox, F. E. (1959). The influence of light reduction on the production, persistence and chemical composition of Coast Bermuda Grass *Cynodon dactylon*. *Agron. J.* **51**, 537–542

Callaghan, T. V., Millar, A., Powell, D. and Lawson, G. J. (1978). "Carbon as a Renewable Energy Resource in the U.K. — a Conceptual Approach". Study undertaken for Energy Technology Support Unit, Harwell by Institute of Terrestrial Ecology

Carruthers, S. P. (1980). Personal communication

Cooper, J. P. and Breese, E. L. (1971). Plant breeding: forage grasses and legumes. *In* "Potential Crop Production" (Wareing, P. F. and Cooper, J. P., eds). Heinemann, London

Davys, M. N. B. (1974). An introduction to dewatering. *Grass* January, 62–78

Ejunjobi, J. K. (1971). Ulex in New Zealand, *J. Ecol.* **59**, 31

Felker, P. (1979). Mesquite, an all-purpose leguminous arid land tree. *In* "New Agricultural Crops" (Ritchie, G. A., ed). American Association for the Advancement of Science

Forrest, G. I. and Smith, R. A. H. (1975). The productivity of a range of blanket bog vegetation types in the Northern Pennines. *J. Ecol.* **63**, 173–202

Gordon, F. J. (1980). "Nutritive Value of Grass and Fresh Grass Products". Paper given at the Nutrition Society (Scottish Group) Meeting, Ayr

Green Europe (1974). "The European Farm: Proteins, Forage Dewatering has Considerable Potential as Indigenous Source for E.E.C. Further research in U.K. should be given high priority". No. 70, pp. 17–24. Agra (London) Ltd

Hills, L. D. (1976). "Comfrey, Past, Present and Future". Faber and Faber, London

Jackson, E. A. (1976). Brazil's national alcohol programme. *Process Biochem.* **11**, 29–30

Jeffries, R. L. (1972). Aspects of salt marsh ecology with particular reference to inorganic plant nutrition. *In* "The Estuarine Environment" (Barnes, R. S. K. and Green, J., eds). Applied Science Publishers, London

REFERENCES

Kirik, M. (1978). A Canadian views alcohol as a farm fuel. *Agric. Eng.* **59** (5), 13–14

Lawson, G. J., Callaghan, T. V. and Scott, R. (1980). "Natural Vegetation as a Renewable Energy Resource in the U.K.". Study undertaken for Energy Technology Support Unit, Harwell by Institute of Terrestrial Ecology

Mason, C. F. and Bryant, R. J. (1975). Production, nutrient content and decomposition of *Phragmites communis* Trin. and *Typha angustifolia* L., *J. Ecol.* **63**, 71–95

Mulcock, A. P. (1978). Lincoln College, New Zealand, quoted by Spedding *et al.* (1979)

National Academy of Sciences, Washington (1977). "Leucaena, Promising Forage and Tree Crop for the Tropics", by Viatmeyer, N. and Cottorn, B. (Ruskin, F. R., ed.)

Nix, J. (1977). "Farm Management Pocket Book". Farm Business Unit, School of Rural Economics and Related Studies, Wye College, UK

Palz, W. and Chartier, P. (eds). (1980). "Energy from Biomass in Europe". Applied Science Publishers, London

Pearsall, W. H. and Gorham, E. (1956). Production ecology, I. Standing crops of natural vegetation. *Oikos* **7**(11), 193–201

Pirie, N. W. (1966). Leaf protein as human food. *Science* **152**, 1701

Pirie, N. W. (1971). "Leaf Protein. Its Agronomy, Preparation, Quality and Use". IBP Handbook No. 20. Blackwell, Oxford and Edinburgh

Plaskett, L. G. (1980). "The Generation of Gases from Biomass by Anaerobic Digestion and their Practical Utilization as Fuel". Study undertaken for Energy Technology Support Unit, Harwell by Biotechnical Processes Ltd

Slesser, M. and Lewis, C. (1979). "Biological Energy Resources". E. and F. Spon, London

Spedding, C. R. W. 91977). An assessment of the future developments of the processes of green crop fractionation in the U.K. *In* "Green Crop Fractionation", p. 177. British Grassland Society, Occasional Symposium No. 9. Harrogate, Yorks.

Spedding, C. R. W., Bather, D. M. and Shiels, L. A. (1979). "An Assessment of the Potential of U.K. Agriculture for Producing Plant Material for use as an Energy Source". Study undertaken for Energy Technology Support Unit, Harwell by the University of Reading

Spedding, C. R. W., Bather, D. M. and Shiels, L. A. (1980). "An Assessment of the Potential of U.K. Agriculture for Producing Plant Material for use as an Energy Source" (Report on Second year of study). Study undertaken for Energy Technology Support Unit, Harwell by the University of Reading

Stahmann, M. A. (1977). "Anaerobic Fermentation for Coagulation of Plant Juice Protein and Preservation of both the Protein and the Fibrous Residues". Paper given at International Workshop on the Utilization of Agricultural Wastes for Feed and Food, C.E.T.C., Belo Horizonte, Brazil

Stewart, D. J. (1978). "Energy Biogas; Production from Crops at Invermay Energy Farm". New Zealand Ministry of Agriculture and Fisheries

Subba Rao, M. S., Singh, N. and Prasannappa, G. (1967). Preservation of wet leaf protein concentrates. *J. Sci. Fd. Agric.* **18**, 295–298

Wilkins, R. J. (1977). "Green Crop Fractionation". British Grassland Society, Occasional Symposium No. 9. Harrogate, Yorks.

Wilkins, R. J. Heath, S. B., Roberts, W. P. and Foxell, P. R. (1977). A theoretical economic analysis of systems of green crop ractionation. *In* "Green Crop Fractionation", p. 131. British Grassland Society, Occasional Symposium No. 9, Harrogate, Yorks

Chapter 6

Grantham, J. B. (1977). Anticipated competition for available wood fuels in the United States. *In* "Fuels and Energy from Renewable Resources" (Tillmann, D. A., Sarkanen, K. V. and Anderson, L. L., eds). Academic Press, New York

Knight, J. A. H., Hurst, D. L. and Thomas, J. F. (1977). Wood oil from pyrolysis of pine bark in sawdust mixture. *In* "Fuels and Energy from Renewable Resources" (Tillman, D. A., Sarkanen, K. V. and Anderson, L. L., eds). Academic Press, New York

Love, P. and Overend, R. (1978). "Tree Power: An Assessment of the Energy Potential of Forest Biomass in Canada". Report ER 78–1. Department of Energy, Mines and Resources, Canada

Molton, P. M., Demmitt, J. M., Donovan, J. M. and Miller, R. K. (1978). Mechanism of conversion of cellulosic waste to liquid fuels in alkaline solutions. *In* "Energy from Biomass and Wastes" (Klass, D. L. and Waterman, W. W., eds). Institute of Gas Technology, Chicago

Terry, M. C. (1978). Franjon fuel pellets. *In* "Energy from Biomass and Wastes" (Klass, D. L. and Waterman, W. W., eds). Institute of Gas Technology, Chicago

Thorensen, E. (1978). Biocoal as an energy carrier. *In* "Energy from Biomass and Wastes" (Klass, D. L. and Waterman, W. W., eds). Institute of Gas Technology, Chicago

Vos, G. D. (1977). Industrial wood energy conversion. *In* "Fuels and Energy from Renewable Resources" (Tillmann, D. A., Sarkanen, K. V. and Anderson, L. L., eds). Academic Press, New York

Zerbe, J. I. (1977). Conversion of stagnated timber stands to productive sites. *In* "Fuels and Energy from Renewable Resources" (Tillman, D. A., Sarkanen, K. V. and Anderson, L. L., eds). Academic Press, New York

Chapter 7

Bio-Energy Directory (1979). Bio-Energy Council. Washington, D.C.

Eliseo, O. and Mariani, P. E. (1978). The eucalyptus energy farm as a renewable source of fuel. *In* "Energy from Biomass and Wastes" (Klass, D. L. and Waterman, W. W., eds). Institute of Gas Technology, Chicago

Fraser, M. D., Henry, J. F. and Vail, C. W. (1976). Design, operations and economics of the energy plantation. *In* "Sharing the Sun Conference" (Winnipeg). International Solar Energy Society, Cape Carnaveral, Florida

Frissel, M. J., Van Goar, C. P., de Hoop, D. W. and Olie, J. (1978). "Feasibility of Short Rotation Forestry for Energy Purposes". Project E Study Report. E.E.C., Brussels. (Quoted in Palz and Chartier, 1980)

Gibson, N. (1978). The economics of wood biomass. *In* "Energy from Biomass and Wastes" (Klass, D. L. and Waterman, W. W., eds). Institute of Gas Technology, Chicago

Inman, R. E. (1977). "Silviculture Biomass Plantations". Mitre Corporation, Metrek Division, USA

Inman, R. E. and Salo, D. J. (1977). Silviculture energy farms. *In* "Fuels from Biomass". Office of Energy Research. University of Illinois, Urbana

Kemp, C. C. and Szego, G. C. (1974). "The Energy Plantation". Proceedings of the 68th National Meeting of the Division of Fuel Chemicals. American Chemical Society

Palz, W. and Chartier, P. (eds). (1980). "Energy from Biomass in Europe". Applied Science Publishers, London

Schneider, T. R. (1973). The efficiency of photosynthesis as a solar energy converter. In "Energy Conversion". Vol. 13, Pt. 3, p. 77. Pergamon, Oxford.

Sieman, J. R. (1975). "The Production of Solar Ethanol from Australian Forests". SES Report No. 75/5. CSIRO, Canberra, Australia

Siren, G. (1976). In "Short Rotation Forestry". Royal College of Forestry, Stockholm. (Quoted in Palz and Chartier, 1980)

Szego, G. C. and Kemp, G. C. (1973). Energy forests and fuel plantations. Chemtech. May, 275. American Chemical Society

Thorensen, E. (1978). Biocoal as an energy carrier. In "Energy from Biomass and Wastes" (Klass, D. L. and Waterman, W. W., eds). Institute of Gas Technology, Chicago

Chapter 8

Bio-Energy Directory (1978). Bio-Energy Council, Washington, USA

Callaghan, T. V., Millar, A., Powell, D. and Lawson, G. J. (1978). "Carbon as a Renewable Energy Resource in the U.K. − A Conceptual Appraoch". Study undertaken for Energy Technology Support Unit, Harwell by Institute of Terrestrial Ecology

Converse, A. D., Grethlein, H. E., Karandiker, S. and Khurtz, S. (1971). "A Laboratory Study and Economic Analysis for the Acid Hydrolysis of Cellulose in Refure to Sugar and its Fermentation to Alcohol". Final Report to the PHS (US). Grant No. US-00597-02. Thayer School of Engineering, Dartmouth College, Hanover, USA

Department of the Environment (1977). Water Engineering Research and Development Division Technical Note No. 6, June, UK

Elton, G. A. H. (1973). The Common Market and food and nutrition in Britain. In "Nutritional Problems in a Changing World" (Hollingsworth, D. and Russell, M., eds). Applied Science Publishers, London

Europool (1977). "Secondary Materials in Domestic Refuse as Energy Sources". C.E.C. Report. Graham Trotman, London

Gordian Associates Inc. (1977). "Overcoming Institutional Barriers to Solid Waste Utilization as an Energy Source". US Department of Energy, HCP/L−50172−01. Washington, D.C.

Greco, J. R. (1977). Energy recovery from municipal wastes. In "Fuels and Energy from Renewable Resources" (Tillman, D. A., Sarkanen, K. V. and Anderson, L. L., eds). Academic Press, New York

James, S. C. (1978). Methane production, recovery and utilization from landfills. In "Energy from Biomass and Wastes" (Klass, D. L. and Waterman, W. W., eds.). Institute of Gas Technology, Chicago

Meller, F. H. (1968). "Conversion of Organic Solid Wastes in Yeasts − an Economic Evaluation". Report for Bureau of Solid Waste Management. US Department of Public Health Education and Welfare. Rockville, Md. USA

Pike, E. B. and Curds, C. R. (1971). The microbial ecology of the activated sludge process. In "Microbial Aspects of Pollution" (Sykes, G. and Skinner, F. A., eds), p. 123. Academic Press, New York

Schlesinger, M. D. (1977). Energy from waste material − 1977 overview. In

"Fuels and Energy from Renewable Resources" (Tillman, D. A., Sarkanen, K. V. and Anderson, L. L., eds). Academic Press, New York

Sinclair, H. M. and Hollingsworth, D. F. (1969). "Hutchinson's Food and the Principles of Nutrition", p. 60. Edward Arnold, London

Stearns, R. P. et al. (1978). Recovery and utilization of methane gas from a sanitary landfill. In "Energy from Biomass and Wastes" (Klass, D. L. and Waterman, W. W., eds). Institute of Gas Technology, Chicago

Wheatley, B. I. and Ader, G. (1979). "Conversion of Biomass to Fuels by Anaerobic Digestion". Energy Technology Support Unit, Harwell, UK

Wright, I. J. et al. (1978). An integrated utility approach to anaerobic digestion of refuse, for electrical power production. In "Energy from Biomass and Wastes" (Klass, D. L. and Waterman, W. W., eds). Institute of Gas Technology, Chicago

Chapter 9

Bach, S. D. and Josselyns, M. N. (1978). Mass blooms of the alga *Cladophora* in Bermuda. *Marine Pollution Bulletin* 9, No. 2

Beneman, J. R. (1978). Recent developments in hydrogen production in microalgae. In "Energy from Biomass and Wastes" (Klass, D. L. and Waterman, W. W., eds). Institute of Gas Technology, Chicago

Bio-Energy Directory (1979). The Bio-Energy Council. Washington, USA

Bryce, A. J. (1978). A review of the energy from marine biomass program. In "Energy from Biomass and Wastes" (Klass, D. L. and Waterman, W. W., eds). Institute of Gas Technology, Chicago

Chapman, V. J. (1970). "Seaweeds and their Uses". Methuen, London

Chin, K. K. and Goh, I. N. (1978). The bio-conversion of solar energy methane production through water hyacinth. In "Energy from Biomass and Wastes" (Klass, D. L. and Waterman, W. W., eds). Institute of Gas Technology, Chicago

Chynoweth, D. P., Klass, D. A. and Ghosh, S. (1978). Biomethanation of giant brown kelp, *Macryocystis pyrifera*. In "Energy from Biomass and Wastes" (Klass, D. L. and Waterman, W. W., eds). Institute of Gas Technology, Chicago

Clendening, K. A. (1917). Organic productivity in kelp areas. In "The Biology of Giant Kelp Beds (*Macrocystis*) in California". (North, W. J., ed). Cramer, USA

Goldman, J. C. and Ryther, J. H. (1977). Mass production of algae: bio-engineering aspects. In "Biological Solar Energy Conversion" (Mitsui, A. et al., eds). Academic Press, New York

Jackson, G. A. (1977). Biological constraints on seaweed culture. In "Biological Solar Energy Conversion" (Mitsui, A. et al., eds). Academic Press, New York

Michanek, G. (1975). "Seaweed Resources of the Ocean". FAO Fisheries Technical Paper No. 138. FAO, Rome

Mitsui, A. and Kamazawa, S. (1977). Hydrogen production by marine photosynthetic organisms as a potential energy resource. In "Biological Solar Energy Conversion". (Mitsui, A. et al., eds). Academic Press, New York

Naylor, J. (1976). "Production Trade and Utilization of Seaweeds and Seaweed Products". FAO Fisheries Technical Paper No. 159. FAO, Rome

Newbold, P. J. (1971). Comparative production of ecosystems. In "Potential Crop Production" (Wareing, P. F. and Cooper, J. P., eds). Heinemann, London

North, W. J. (1977). The ocean as a setting for biomass production. *In* "Fuels from Biomass". Office of Energy Research. University of Illinois, Urbana

Oswald, W. J. and Beneman, J. R. (1977). A critical analysis of bio-conversion with microalgae. *In* "Biological Solar Energy Conversion" (Mitsui, A. *et al.*, eds). Academic Press, New York

Oswald, W. J., Beneman, J. R. and Koopman, B. L. (1977). Production of biomass from freshwater aquatic systems — concepts of large-scale bio-conversion systems using microalgae. *In* "Fuels from Biomass". Office of Energy Research. University of Illinois, Urbana

Palz, W. and Chartier, P. (eds). (1980). "Energy from Biomass in Europe". Applied Science Publishers, London

Roels, C. A., Lawrence, S., Farmer, M. W. and Van Hemelryck, L. (1976). Organic production potential of artificial upwelling marine culture. *In* "Microbial Energy Conversion". Erich Goltze, Gottingen

Ryther, J. H., Lapointe, B. E., Stenberg, R. W. and Williams, L. D. (1977). Cultivation of seaweed as a biomass source for energy. *In* "Fuels from Biomass". Office of Energy Research. University of Illinois, Urbana

Shelef, G., Moraine, R., Meydan, A. and Sandbank, E. (1976). Combined algal production — waste treatment and reclamation systems. *In* "Microbial Energy Conversion". Erich Goltze, Gottingen

Soeder, C. J. (1976). Primary production of biomass in freshwater with respect to microbial energy conversion. *In* "Microbial Energy Conversion". Erich Goltze, Gottingen

Tsudaka, O., Kawahara, T. and Miyachi, S. (1978). Mass culture of Chlorella in Asian countries. *In* "Biological Solar Energy Conversion". (Mitsui, A. *et al.*, eds). Academic Press, New York

Westlake, D. F. (1963). Comparison of plant productivity. *Biol. Rev.* (Cambridge Phil. Soc.), No. 38, 385

Chapter 10

Gary, H. P., Pande, P. C. and Thanvi, K. P. (1980). Designing a suitable biogas plant for India. *In* "Appropriate Technology". Vol. 7, No. 1. Intermediate Technology Publications, London

Gordian Associates Inc. (1977). "Overcoming Institutional Barriers to Solid Waste Utilization as an Energy Source". US Department of Energy. HCP/L–50172–01, Washington, D.C.

Hall, D., Adams, M., Gisby, P. and Rao, K. (1980). Plant power fuels hydrogen production. *New Scientist* **86**, 72

Hammond, A. L. and Metz, W. D. (1977). Solar energy research: making solar energy after the nuclear model. *Science* **197**, 241

Intergroup Consulting Economists Ltd (1976). "Economic Pre-feasibility Study: Large Scale Methanol Fuel Production from Surplus Canadian Biomass". Fisheries and Environment Canada. Policy and Programme Directorate, Ottawa

Klass, D. L. (1974). "A Perpetual Methane Economy. Is it Possible?". Chemtech., p. 161. American Chemical Society

Mitsui, A. and Kamazawa, S. (1977). Hydrogen production by marine photosynthetic organisms as a potential energy resource. *In* "Biological Solar Energy Conversion". (Mitsui, A. *et al.*, eds). Academic Press, New York

Overend, R. (1978). "The Role of R and D in Biomass Energy Development". Department of Energy, Mines and Resources, Canada

REFERENCES

Slesser, M. and Lewis, C. (1979). "Biological Energy Resources". E. and F. Spon, London

Van Buren, A. and Pyle, L. (eds). (1979). "A Chinese Biogas Manual". Intermediate Technology Publications, London

Wilcox, H. A. (1980). Ocean farming prospects and problems. *SPAN* (Shell Chemicals, UK) 23, No. 2

Subject Index

Acetaldehyde, 14
Acetic acid, 15, 24, 35, 40, 107, 119
Acetone, 24, 71
Acid hydrolysis; see Hydrolysis
Acid rain, 186
Activated carbon, 150
Acyclic non-isoprenoid hydrocarbons, 47
Adenosine triphosphate (ATP), 43
Aerobic fermentation, 44
Agar, 165
Alcohol, 26; see also Ethyl alcohol, Methyl alcohol
Alder, 7, 130
  red, 125
  European, 131
Algae, 5, 7, 35, 47, 179
  blue-green, 176
Algin(s), 159, 160, 164, 165
Alkali treatment, 70
Alkaloids, 55
Almond shells, 19, 50
Alter programme, 183
Alum, 170
Amino acids, 153
Ammonia, 17, 37, 41, 47, 70, 77, 84
Ammonium salts, 43
Ammonium sulphate, 166
Amylase, 43, 44
Anabaena cylindrica, 47, 173
Anaerobic digestion, 20, *34*–46, 63, 65, 67, 68, 73, 76–78, 81, 83–85, 92, 93, 104, 157, 176
  of algae, 173
  of MSW, 144, *147*–148, 153
  of seaweed, 164
  of sewage sludge, *139*–140
Animal excreta, *73*–87; see also Livestock wastes
Anthraquinone disulphonic acid, 41
Antibiotics, 71

Aphanothece sp., 169
Aqueous distillate, 25
Ascophyllum spp., 8, 160, 165
Ash, 17, 21, 23–29, 31, 37, 63, 65, 70, 71, 73, 74, 78, 90, 112, 145, 146, 159, 160
Ash (tree), 113
Aspartate, 5
Aspen, 112
Aspergillus niger, 44
Autothermal gasification, 151
Azeotrope, 44

β-Terpineol, 16
Bacteria, 138, 148, 173, 179
  facultative anaerobes, 35
  methanogenic bacteria, 35, 37, 148
  rhodophytic bacteria, 4
Bagasse, 20, 22, 23, 44, 50, 51, 71
Ball milling, 30
Bamboo, 71
Barley, 6, 49, 51, 52, 53, 63, 69
Beans, 50, 51, *55*, 56, 60, 63, 96
Beech, 112
Beet; see Sugar beet
Benzene rings, 13
Benzol, 178, 184
Berula sp., 7, 156
Bilirubin and biliverdin, 74
Birch, 7, 113
  white, 112
Biogas, 40, 41, 42, 67, 73, 78, 81, 83, 86, 104, 136, 149, 171, 184
Biocoal, 117
Biological conversion processes, 17, 18, 19, *34*–47
Biological Oxygen Demand (BOD), 138
'Biomass Society' (the), 185
Biophotolysis, 47, 173, 179
Bioplex concept, 85

Bioproductivity, 8
Biotechnology revolution, 175
Boiler(s)
  efficiency of, 20
  shell boilers, 22
  water tube boilers, 22
  wood burning, 111, 116
Bracken (*Pteridium aquilinum*), 7, 99, 100, 101, 109, 110
Brassicas, 51, *56*, 60
Briquettes, briquetting
  of straw, 68
  of wood, 21, 118
Broccoli, 51, 56
Brussel sprouts, 51, 56, 60, 63, 105
Butanol, 71
Butyric acid, 35

Cabbage, 51, 56, 63
Catechols, 71
Calcium, 37
Calcium acetate, 166
Carbon (C), 1, 5, 8, 11, 12, 13, 14, 15, 28, 29, 66, 77, 89, 117, 121, 146, 159, 168, 173, 176
Carbohydrate(s), 1, 3, 4, 13, 15, 25, 41, 47, 90, 92, 159, 160
Carbonisation, *25*
Carbon dioxide (CO_2), 4, 11, 15, 18, 19, 24, 25, 26, 34, 35, 40, 41, 117–120, 148, 149, 150, 151, 168, 171, 172, 172, 176
  atmospheric, 180
  balance, 124
Carbon monoxide (CO), 14, 19, 25, 27, 28, 34, 117–120
Carbon/nitrogen ratio, 40
Carboxyl group, 13
Carex, 7
Carotenes, 13, 16
Carrageenan, 161, 163, 165
Carrots, 51
Cassava, 44, 45, 50, 90, 93, 95, 101, 178, 181, 184
Catalysts, 33, 34
Catch crops, 62, 91, *95–99*, 105, 110
Cauliflower, 51, 52, 56, 63
Cellulose, 12, 13, 35, 38, 45, 46, 65, 68, 71, 74, 75, 90, 112, 120, 121, 147, 153, 159, 160, 164, 178

Centrifugation, 44, 86
Ceratophyllum demersum, 7, 156
Cetyl alcohol, 16
Char, 25, 26, 27, 30, 115
Charcoal, 24, 117, 118, 119
Chicory, 4
Chipping, 29, 30, 116; *see also* Hog fuel
Chlorella sp., 167, 168
Chloride content, 158, 164
Chlorinated hydrocarbons, 37, 140
Chlorine, 146, 151
Chlorophyll, 4, 179
Cholesterol, 13
Chondrus crispus, 161, 163, 165
Citric acid, 71
Cladophera prolifer, 167
Cloning, 127, 132
Clover, 65, 91–94
Coal, 1, 11, 17, 20, 23, 29, 30, 34, 67, 71, 117, 128, 145, 177, 182
  anthracite, 12
  bituminous coal, 12, 146
  brown coal, 21
Cobalt, 37
Cocceryl alcohol, 10
Coconut shells, 50
Coke, 17, 27, 28
Colza, 50, 52
Combustion, 11, 17, *19–23*
  of animal wastes, 73, 75, 86
  of crop wastes, *49–51*
  of MSW, *144*–145, 152, 185
  of natural vegetation, 89
  of wood, *116*–118, 182, 183
Comminution, 19
Common bent grass (*Agrostis tenuis*), 109
Comfrey (*Symphytum asperrimum*), 102
Cotton linters, 71
Cottonwood, 130
Crop fractionation, 68, 72, 194, 195, 106, 107, 108, 109
Crop wastes, residues, 1, *49–72*, 103, 105, 121, 181
Crude fibre, 63, 64, 65
Cutin, 13
Cyclone furnaces, 22, 32
Cynodon dactylon, 100
Cyperus papyrus, 155

SUBJECT INDEX

Dehydration, 17, 19, 29, 61, 62, 68, 75, 86, 106
Denaturation, 178
Densification, 20, 21, 29, 30, 58
Detergents, 37, 140
Diesel fuel, 184
Digesters, 34, 35, 37, 41, 42, 68, 76, 78, 81, 82, 85, 94, 139, 140, 141, 171, 173, 184
  contact digesters, 39
  high rate slurry digesters, 38, 39, 81
  packed bed reactors, 39
  plug flow, 36, 39
  solid form batch digesters, 39
  unstirred slurry digesters, 39
Digestion–residence, retention, detention times, 38, 42, 83, 108, 137, 138
Digestible solids, 37
Distilleries/Distillation plant, 44, 182
Dunaliella sp., 169

Ecofuel, 30, 146
Eichhornia crassipes, 155, 156, 157
Electricity generation, 17, 25, 42, 67, 78, 83, 93, 116, 117, 126, 131, 144, 147, 148, 182, 183
Electrolysis, 179
Electrostatic precipitation, 32, 145
Energy accounting, 180
  balance, 129
  budget, 18
  conversion efficiency, 18
Embden-Meyerhof Pathway, 43
ENFOR Programme, 183
Enzymic hydrolysis, 46, 173, 179
Epilobium sp., 100
Ethane, 14, 120
Ethanol; see Ethyl alcohol
Ethyl alcohol (ethanol), 13, 14, 18, 35, *43*–46, 71, 90, 92, 94, 104, 107, 114, 120, 153, 177, *178*–179, 180, 181, 182, 184
Ethylene, 71
Ethylene glycol, 15
Euchina sp., 161, 165
Eucalyptus spp., 123, 125, 126, 130
Euphorbia spp., 104

Fat(s), 12, 41, 47, 159
Fatty acids, 13, 15, 35

Fatty oils, 13, 47, 176
Fermentation, 2, *43*–46, 47, 107, 153, 178, 181
Fermenters, 43; *see also* Digesters
Ferric chloride, 170
Firewood, 1, *111*–113
  reserves, 123
Fischer-Tropsch process, 34
Flue gas, 20
Fluidised bed, 21–23, 29, 31, 32, 145
Fodder beet, 93, 98
  crops, 52, 91, 92
Forest energy plantations, 3; *see also* Short rotation forestry
Forest industry wastes, 112–114, 122
Forest wastes/residues, 112, 113, 122
Formaldehyde, 14
Formic acid, 13, 14
Franjon pellets, 118
Fructose, 44, 153
Fruit crop wastes, 57
Fucaloids, 165
Fucoidin, 159, 160
Fucus, Fucoids (wracks), 158, 160
Fuel pellets, 29, 92, 118
Furan resins, 71
Furcellaria sp., 160, 165
Furfural, 71
Furfuryl alcohol, 71
Furnaces
  cyclone furnaces, 22
  straw furnaces, 60, 62, 66, 67

Gas cleaning, 32
  scrubbing, 145
Gasification, 17, 27, 29, 31, 32, 68, 92, 183
  air gasification, 26–28, *145*–148
  air free gasification, 151
  of MSW, *151*–152
  oxygen gasification, 26–28
  water and steam gasification, 28
  of wood, 118–121, 176
Gasifiers, 27, 31
  co-current moving bed type, 151
  cross-flow type, 31
  cyclone reactors, 32
  downdraft type, 31
  fixed bed reactors, 31
  fluidised bed type, 32
  horizontal moving bed type, 32

Gasifiers (cont.)
  rotary kiln, 32
Gasifier gas, *32*–34
Gasohol, 120, *178*–179, 181, 182, 184, 185
Gasoline, 26, 34, 177, 178, 181; see also Petrol
Gelidium sp., 165
Gigartina sp., 165
Glasshouse heating, 67
Glucose, 13, 45–47, 71, 153
Glycolic acid, 13, 15
Glyoxal, 15
Glyoxylic acid, 15
Gobar Gas Institute
  design, 39
  units, 184
Gorse (*Ulex europaeus*), 100
Gracilaria sp., 8, 161, 163, 165
Grain, 44, 49, 52, 67, 73, 178, 182
  drying, 67
  yield, 49, 52, 53
Grain/straw ratios, 52, 53
Grand Fir, 7
Grass, 12, 65, 74, 77, 92, 94, 105
  yields, 103
Gross Primary Production (GPP), 4
Gunnera sp., 100, 101
Gut flora, 74

Heat exchangers, 38
Heat of formation, 13
Heather (*Calluna vulgaris*), 109 110
Hemicellulose, 35, 45, 68, 71, 112, 121
Hemp, 50
Heracleum, 100
Hevea braziliesis, 104
Hexose, 47
High rate ponds, 169–171
Hog fuel, hogging, 116–118
Holmes-Stretford process, 41
Hormones, 129
Humus, humus levels, 49–51, 68, 73, 89
Hydrasposal, 146
Hydrilla verticilliata, 157
Hydro-electric power, 1
Hydrogasification, 34, 120
Hydrogen, 1, 11–13, 18, 24–29, 32, 34, 35, 41, 47, 119–121, 146,

Hydrogen (cont.)
  151, 152, 159, 178, *179*
  fuel, 12, 173, 179
  sulphide, 41, 119, 149
Hydrogenase, 173
Hydrogenation, 17, 151
Hydrolysis, 38, 43–46, 71, 75, 121, *153*, 178

Impatiens sp., 100
Incinerators
  controlled combustion type, 23
  water wall type, 144–146
Industrial effluent, 135–137
Insolation; see Solar radiation
Internal combustion engines, 178
Inulin, 44
Irideae cordalia, 163
Isoprene, 16
  subunits, 13
Isoprenoid hydrocarbons, 13, 104
ITE land classes, 97

Jerusalem artichoke, 44, 92
Jute, 50

Kale, 93, 98
Kelp(s), 161, 165, 166, 177; see also *Macrocystis*
Kilns, 119
Kraft process, 117

Lactic acid, 35, 71
Laminaria, 8, 158
Laminarin, 159, 160, 165
Land-fill gas, *148*–150, 176
Lemna minor, 157
Leguminosae, 176
Leucaena sp., 194
Lignin and ligno-cellulose, 12, 13, 15, 35, 37, 45, 46, 71, 73–75, 112, 176
Limomium sp., 110
Limonene, 16
Linoleic acid, 16
Lipids, 35
Livestock wastes, 2, 27, 33, 35–37, 40–42, 69, *73*–87, 193
Locust (tree), 130
Lucerne, 91–94, 105

SUBJECT INDEX

Macrocystis pyrifera (giant kelp), 8, 158, 159, 162–165
Macrophytic seaweed, algae, 8, 155, *157–167*
Magnesium, 37
Maize, 5, 6, 51, 57, 59, 93, 96, 183
  stover, 50, 52
Malate, 5
Manioc; *see* Cassava
Mannitol, 159, 160
Manure; *see* Livestock waste
Maple, red, 112, 130
Mariculture on land, *169–174*
Meat, 12
Melle-Boinot method, 44
Mesophilic reaction, 36
Mesquite; *see Prosopis*
Metabolisable energy content (ME), 69, 70
Methane, 12, 14, 18, 19, 24–29, 32, 34–37, 40, 47, 75, 85, 101, 108, 119, 121, 139, 142, 147, 148, 151, 153, 172, 173, *176–177*
Methane economy (the), 176
Methanobacterium sp., 35
Methanol; *see* Methyl alcohol
Methyl alcohol (methanol), 13, 14, 18, 19, 24, 26, 32, *33–34*, 47, 90, 114, 119, *120–*121, 144, 152, *177–*178, 181, 183
Methylene units, 12
Microalgae, 2, 7, 8, 155, 161, *167–*174
Microbial actions, 19, 34, 35 40, 47, 71, 90
  protein, 85
Mobil process, 34
Molasses, 44
Molecular sieves, 150
Multi-stage reactors, 40
Municipal Solid Waste (MSW), 1, 30, 33, 40, 45, 68, 112, 121, 131, 135, *141–*153, 169, 176, 185
  air and water classification of, *145–*148
Mustard, 65

Napier grass (*Pennesetum purpureum*), 6
Naphtha, 165

Natural gas, 11, 32, 120, 128, 148, 149, 176
Net Primary Production (NPP), 4
Nitrogen (N), 17, 26, 40, 41, 43, 47, 66, 74, 77, 84, 86. 87, 93, 94, 119–121, 146, 148, 149, 159, 171, 172
  oxides of, 117, 118, 131, 133
n-tria-contanol, 16
Nuclear fuel, 17
  energy, 179, 181
  fusion, 186
Nucleic acid, 74

Oar weed; *see Laminaria*
Oats, 51–*52*, 63, 93
Ocean thermal energy, 180
Octane rating, 12, 178
Olefins, 26
Oleic acid, 16
Olive processing waste, 52
Oocystis solitaria, 170
OPEC, 181
Oxalic acid, 13, 15
Oxidation, 18, 19, 151
Oxygen, 1, 4, 11, 12, 17–19, 24–27, 29, 35, 146, 151, 152, 172, 173

Palm oil wastes, 52
Paraffin(s), 28, 34, 166
Paraffin alcohols, 13
Pasteurisation, 4
Pathogens, 68
Peanuts, 12
Peas, 50, 51, *55*, 56, 60, 63, 96, 98
Peat, 23
Pentoses, 71
Pesticides, 71
Petroleum, petrol, 24, 34, 43, 114, 178, 179, 182, 184, 186
Pharmaceuticals, 71
Phenolic chemicals, 24, 71
Phenylpropane, 13
Phosphate (PO_4), 17, 66, 84, 87, 94
Phosphoglyceric acid (PGA), 5
Photocoins sp., 170
Photolysis, 173
Photosynthesis, 1, 3, 4, 5, 47, 93, 183
  C_3 and C_4 routes, 5, 168
Photosynthetic efficiency, rate, 1, 5, 6, 96, 99

Photosynthetically active radiation, 3
Photo-voltaic systems, 180
Phragmites sp., 7, 100, 110, 155, 156
Phyllopharae sp., 158
Phytoplankton, 8, 167
Pine, 112, 113, 131
Plantation crops, *99*–104
Plant culture, 47
Polygonum sp., 100, 102, 105
Polygonum cuspidatum, 101
Polymerisation, 26
Polysaccharides, 35, 43, 44
Polythene, 71
Polyvinyl chloride (PVC), 71
Poplars, 7, 113, 123, 125, 130
Potatoes, 5, 6, 44, 51, *54*–55, 57, 62, 90, 92–94, 96, 102
Potash (K_2O), 84, 94
Potassium (K), 17, 37, 66, 87, 99, 159
Power alcohol, 178; *see also* Gasohol
Proalcool, 181, 182
Process plant waste, 112
Producer gas, 17, 27, 119, 120
Prosopis sp., 104, 130, 131
Protein(s), 12, 17, 35, 40, 41, 62, 63, 65, 68, 74, 85, 86, 105–108, 159, 160, 168, 169
Protozoa, 138
PUROX process, 120, 152
Purple moor-grass (*Molinia caerulea*), 109
Pyroligneous acid, 24, 119
Pyrolytic oil, pyrolysis oil, 1, 24, 26, 68, *121*, 144, 151, 152
Pyrolysis, 21, *25*–27, 68, 92, 116, 176, 181
 of algae, 173
 of MSW, 142, 145, *151*–152
 of seaweed, 166
 of wood, *118*–121

Quarter Acre Module (QAM), 163

Radish, 91, 92, 98
Raney nickel catalyst, 33
Rape, 50, 52, 63, 96, 98
Rapeseed, 12
Ranunculus sp., 7, 156
Reduction, 19, 28

Refuse-derived solid fuel (RDSF), 144, *145*–147, 185
Rice, 50, 51, 57, 72
Rubber, 10, 13, 153
Ruminants, 74
Rye, 50–53, 63, 91, 93
Ryegrass, 91, 93, 94, 99

Saccharification, 44
Saccharomyces cerevisiae, 41
Sagitaria sp., 156
Sargassum sp., 8, 158
Savoys, 51, 56
Sawdust, 2, 26
Scenedesmus spp., 8, 168, 170
Scots pine, 6
Seaweed, 2, 8
Sewage, 35, 37, 41, *135*–141, 171–173, 176
Sewage treatment
 activated sludge process, 137
 oxidation, 137, 138, 171
 trickling filter method, 137, 138
Shift reaction, 33
Short Rotation Forestry (SRF), *123*–133, 184
Silicon gel, 150
Single cell protein, 71
Sisal, 50
Slag, 27, 28
Slurry separation, 86
Sodium, 37
 hydroxide, 70
Soil fertility, 17
Solar energy, 1, 3, 4, 6, 90, 190
 constant, 3,
 insolation rates, 126, 127
 conversion efficiencies, 2–7, 124–127, 130, 132, 173
 radiation, 3, 124, 186
 zenith, 3
Sorbitol, 71
Sorghum, 178, 183, 184
Soya beans, 181
Spartina sp., 100, 110
Sphagnum moss, 110
Spirulina sp., 167–169
Spruce, 112
Starch, 12, 13, 43, 44, 90
 crops, 178, 181, 182

SUBJECT INDEX

Steam gasification, 28
Stearic acid, 16
Steroids, 13
Stinging nettle (*Urtica* sp.), 100
Strain selection, 47
Straw, cereal straw, 2, 3, 12, 15, 20, 21, 23, 24, 30, 40, 42, 49, 50, 51, *52–53*, 58, 59, 62, 63, 66–68, *69*–71, 84, 85, 80, 81, 110
  alkali treatment of, 70
  ammoniation of, 70
  yields, 49, 52, 53
Suberin, 13
Sucrose, 153
Sugar(s), 43, 45, 46, 53, 90, 92, 93, 95, 104, 107, 121, 153, 178
  beet, 5, 6, 51, *53*–54, 57, 63, 65, 66, 71, 90, 91, 92, 93, 94, 95, 96, 99, 102
  cane, 5, 20, 43, 44, 50, 51, 71, 90, 95, 178, 181, 184
Sulphide, 37
Sulphite liquor, 21
Sulphonates, 37
Sulphur, 29, 32, 41, 119, 121, 151, 159, 166
  dioxide (SO_2), 41, 176
Sulphuric acid, 18, 45, 121, 153
Surfactants, 71
Sweet gum, 130
Sycamore, 113, 130
  American sycamore, 125
Synthesis gas, 177
Synthetic rubber, 71

Tar, 27, 32, 119, 120
Tea, 50
Terpenes and terpenoids, 13, 16
Textile chemicals, 71
Thermal conversion/processing, 17, 19, *23*–34, 50, 68, 111, 126
Thermal decomposition, 18
Thermophilic reaction, 36, 38
Town gas, 17
Trace elements, 17
Transport costs, economics, 58, 59, 61, 79, 94, 95, 122, 146
Typha sp., 7, 110, 155, 156

Urea, 74
Uric acid, 74

Vanillin, 71
Vapour cracking, 26
Vegetable crop wastes, 51, *57*, 68, 110
Vetch, 98
Visible light, 2
Vitamins, 43
Vitamin B_{12}, 37
Vitrinite, 29
Volatilisation, 17, 18
Vortex principle, 116

Waste-Derived Fuel (WDF), 145
Water gas, 17, 28
Water hyacinth; *see Eichhornia crassipes*
Wave power/energy, 1, 187
Waxes, 13, 16
Wheat, 6, 12, 49, 51–53, 63, 69
  yields, 132
Whole crop harvesting, 61, 62, 97, 98
Whole tree harvesting, 112, 113, 122
Wind power, 1, 187
Wood, firewood, wood waste, 1, 2, 12, 13, 15, 23, 24, 30, 50, 75, 80, *111*–122, 176, 181, 182, 183
  distillation, 119
  gas, *119*–120, 176, 177, 183
  industry, *113*–116
  oil, 121
  pulp, 70, 71
  tar, 24
Woody wastes in agriculture, 57
Wracks; *see Fucus*

Xylose, 71

Yeast, 43, 71

Zeolites, 41
Zero grazing, 77
Zostera marina (eel grass), 8, 157, 164